Dr. Jensen's
GUIDE TO BODY CHEMISTRY & NUTRITION

Dr. Jensen's GUIDE TO BODY CHEMISTRY & NUTRITION

Bernard Jensen, D.C., Ph.D.
Clinical Nutritionist

KEATS PUBLISHING
LOS ANGELES
NTC/Contemporary Publishing Group

The purpose of this book is to educate. It is sold with the understanding that the publisher and author shall have neither liability nor responsibility for any injury caused or alleged to be caused directly or indirectly by the information contained in this book. While every effort has been made to ensure its accuracy, the book's contents should not be construed as medical advice. Each person's health needs are unique. To obtain recommendations appropriate to your particular situation, please consult a qualified health-care provider.

Library of Congress Cataloging-in-Publication Data

Jensen, Bernard, 1908–
[Guide to body chemistry & nutrition]
Dr. Jensen's guide to body chemistry & nutrition / Bernard Jensen.
p. cm.
Includes index.
ISBN 0-658-00277-5
1. Minerals in human nutrition. I. Title: Doctor Jensen's guide to body chemistry and nutrition. II. Title.
QP533 .J46 2000
612.3'9—dc21 00-035710

Design by Andrea Reider

Published by Keats Publishing
A division of NTC/Contemporary Publishing Group, Inc.
4255 West Touhy Avenue, Lincolnwood, Illinois 60712, U.S.A.

International Standard Book Number: 0-658-00277-5

20 21 22 23 24 25 26 LHN 20 19 18

CONTENTS

INTRODUCTION

Knowing body chemistry is important because people's lives and health are important. To place this on a personal level, *your* life and health are important. The reason I have devoted my life to the healing arts and teaching people how to live healthy lives is because I truly feel people—like you—are important. If you want to have a healthy life, you have to learn how to go about it. I have invested seventy years of my life learning how to take care of patients at my "live-in" health ranches, following what is called "the nature cure" philosophy. While my patients were getting well, I gave them lessons in right living, using knowledge I had been given by my teachers and by personal experiences and travels.

I met my most important teacher, V. G. Rocine, a few years after I graduated from West Coast Chiropractic College in Oakland, California. Rocine was a Norwegian homeopathic doctor, and his lecture was about food. Food chemistry was new at that time, pioneered by European food chemists such as Dr. Ragnar Berg of Sweden and Dr. Georg Koenig of Germany. Rocine's lecture touched something deep inside me.

I knew I had to have this food knowledge for myself and for my patients.

Rocine taught that our bodies are made of the "dust of the earth," the chemical elements that make up the soil of our planet. He pointed out that unless we know and use foods that have the right chemical elements needed to sustain health, we are hopelessly unable to resist disease. I felt like he was talking to me. I had been raised on coffee and Danish pastry, and my health had already suffered one major crisis as a result. I needed to get this food knowledge that he was offering.

I studied with Rocine for a while, read all of his books (now mostly out of print), and attended many of his lectures in subsequent years. Over the years, he researched the symptoms of dietary deficiencies and excesses of what he considered the sixteen most important chemical elements. He was the first person I know who believed in the "window theory" of limitations of food minerals. That is, if your intake is below a certain amount, you experience deficiency symptoms, and if your intake is above a certain amount, you experience toxic effects. (The latter is now known to be only true of some nutrients, not all, but the "window theory" is widely accepted for most nutrients.) You have to take in the right amount of minerals, within upper and lower limits, to get the most good from them.

Rocine urged me to use foods to heal myself and my patients, and I did. My experience with patients and contact with other doctors helped me to build on the foundation I learned from Rocine. But he was the one who sparked the fire that got me started.

I love working in the healing arts. I love seeing people get well and leaving their troubles behind. You have to learn how to be well by cooperating with nature. Disease and loss of good health are not accidental. You have to violate nature's laws to lose your health and develop a disease. You have to eat, drink, and think yourself into a disease. You have to work hard to break down some parts of your body. If you want out of a disease, you have to work your way out of it just as you worked your way into it. You start to reverse this disease way of life by learning a right way of living. You have to learn a few important things about how your body works and how to meet its needs for foods, exercise, fresh air, clean water, rest, and recreation.

To learn how to get healthy and how to stay healthy, you'll have to learn how food chemistry relates to body chemistry. Chemistry is simply the knowledge of how atoms and molecules make up the structure of matter and how matter of one kind becomes changed into matter of another kind. Body chemistry teaches you how and why you need certain amounts of nutrients. Food chemistry teaches you the best food sources for those nutrients.

Most people don't have the slightest idea how hard their bodies work to keep them alive and well. There are over fifty thousand proteins being manufactured and used in your body in the normal course of ongoing life, billions of neurons firing in your brain and nervous system all the time, millions of new red blood cells being made each second to replace the millions that are worn out, billions of white blood cells destroying harmful microorganisms and cleaning up debris in your body, all of this and more below your conscious awareness. People say, "I need a miracle," but the Chinese say, "You *are* a miracle." The Chinese were treating goiters nutritionally five thousand years ago, bringing edible seaweed, dried fish, and burnt sponges over a thousand miles inland to treat those who had goiter due to iodine deficiency. They didn't have the slightest clue what caused the swollen glands in people's throats or why foods from the sea cured the swelling. All they needed to know was what to do for patients with bulging throats.

Your doctor is not responsible for your health. He or she only becomes responsible for that part of your body that isn't working right when you show up at the clinic asking for help. Your doctor may or may not be able to help you. Doctors are often taught, "Nature does the healing, but you be sure to take credit for it and send a bill to your patient." What I'm saying is, "You are responsible for your health."

You, more than anyone else, know what you do every day that affects your body. You are a unique human being, and only you can make the adjustments in your diet and lifestyle necessary to be the healthiest and best person you can be. You can't put that responsibility onto your doctor, who can't do it for you. If you are motivated to build the best health you can and all you lack is the right knowledge, that can be provided. I learned the hard way.

I developed a severe lung infection after I graduated from chiropractic school, and I didn't know what to do. I knew you couldn't fix a lung infection with chiropractic, so I went to a regular doctor. He said, "There's nothing I can do for you. Just go home and go to bed and see if the infection will heal." I was shocked. He didn't know what to do either! Of course, that was in the days before antibiotics. Lack of knowledge in the face of a terrible disease or infection is very frightening. I wasn't about to lay in bed until I found out whether I was going to sink or swim, so I began to search for someone who really knew what to do. This was before I met Dr. Rocine.

I found a Seventh-Day Adventist doctor who knew foods, and who told me that my poor nutrition history undoubtedly contributed to my infection. He put me on a diet with lots of fresh fruits and vegetables, especially green vegetables, and I began to pull out of my health slump. I also started breathing exercises designed by Thomas Gaines, a physical education teacher in New York City, and that helped even more. I recovered from the infection, and not long afterward attended my first Rocine lecture.

In 1938, I wrote to Rocine about my beginning nutrition work with patients, and in his letter of response he wrote, "Go on with your work, Dr. Jensen. Men like you are needed by the millions in this world. People, as a general rule, fail to study diets." I received one last letter from him just before he died, in which he responded to my news about starting a sanitarium in the mountains of Southern California. He wrote, "You have the right idea of having a health [sanitarium] in nature, where nature cures when a cure is possible." Dr. Victor G. Rocine died on February 25, 1943. Before he died, he called me one of his best students, and I felt honored by his assessment.

Dr. Rocine would have been surprised and delighted if he could have traveled with me to the Hunza Valley of Pakistan in the 1950s where I encountered men who lived to be 120, 130, and 140 years old. They ate simple foods raised in their own gardens and fields and were isolated from the rest of the world eleven months of the year as snow and ice forced closure of the mountain passes. Most of them had every tooth in their heads, good hearing and vision, and clear memories both of recent events and things that happened when they were very young. They still worked in the fields every day of the growing season. Most of the older women spent the greatest part of their time indoors, helping their children's families, and doing domestic chores. They seldom lived past their nineties, for reasons perhaps related to their indoor work. The Hunza people had no doctors, hospitals, drugstores, jails, prisons, police, or modern conveniences (except for the Mir, who governed the Hunza people and lived in a modern, but modest palace). Rocine would have felt vindicated in his teaching that a simple, natural diet of nutritious food is the primary secret to good health and longevity.

What made their diet so nutritious? They irrigated their fields terraced into the mountainsides with mineral-rich glacier water from the high mountains surrounding their valley. Their food crops were as nutrient rich as foods can get. The people also drank the water, which was cloudy from the high mineral content. They cultivated fruit orchards, vegetables, and cereal grains, and used a little meat and honey in their diets. Because of the isolation of their valley, they had no supermarkets, convenience stores, or bars—no place to buy alcoholic beverages, cigarettes, soft drinks, sugar, coffee, or processed foods of any kind. There was nothing to war against their health, nothing to promote favorable disease conditions.

The Hunza Valley residents were an inspiration to me of what can happen when people live a simple, healthy life using only whole, pure, fresh, and natural foods, getting enough exercise, breathing clean air, and drinking pure water—all in a social context of peace and harmony. Rocine would have loved to see the people of that valley.

The Hunza people didn't need to know food chemistry or body chemistry, but we do because we live in a wealthy nation in which health risks are common and health wisdom is uncommon. Only by educating and dedicating ourselves to make the right food choices and the right lifestyle choices, and by living in harmony with nature and other people, can we make the most of the new millenium by embarking upon a better, happier, and healthier life than the generations that preceded us.

CHAPTER 1

THE TWENTY-ONE CHEMICAL ELEMENTS PEOPLE NEED

I have studied the twenty-one chemical elements that are essential to the health and well-being of all people, and I have supervised the dietary regimens of thousands of individuals in various stages of health and disease at my health ranches in California. I have observed firsthand what happens when you change from poor food habits to a balanced, nutrient-rich food regimen. I have seen miracles take place in the health of individuals whose doctors had given up on them after trying every conventional medical treatment they knew. The miracles took place after patients began receiving the nutritional support their bodies needed.

Our vital organs, glands, bones, ligaments, muscles, and other tissues of the body each have their own special nutritional needs (see Table 1.1). People's nutrient needs may be similar in many respects, but each person differs regarding the exact amounts of nutrients needed and may require special supplements to catch up on long-term deficiencies in certain organs or tissues. Each tissue type is made up of millions of tiny cells that act like microscopic factories, taking in raw materials that are used partly for food, partly for energy production, and partly to manufacture substances needed by other cells.

If we had a powerful enough microscope to see a single cell in operation, we could find out what happens when there is a shortage of one or two chemical elements needed by that cell. The first thing we would see is that the cell can't function adequately without all the nutrients it needs. It has to produce less of what it is supposed to make for other cells, or else the sub-

Table 1.1. **Chemical Needs of Body Organs**

Tissue Type	*Nutrient*
Adrenals	Zinc
Blood	Iron
Bowel	Magnesium
Brain/Nerves	Phosphorus/Oxygen
Heart	Potassium/Magnesium
Kidneys	Chlorine
Liver	Iron/Sulfur
Lungs	Silicon
Muscles	Potassium
Nails/Hair	Silicon
Pituitary	Phosphorus
Skin	Silicon/Sulfur
Spleen	Copper/Chlorine
Stomach	Sodium/Chlorine
Teeth/Bones	Calcium/Fluoride
Thyroid	Iodine

stance it is making will be abnormal in its structure and quality. If a cookie recipe calls for flour, eggs, honey, vegetable oil, and raisins, I can guarantee you will be disappointed if you leave out the flour or any of the other ingredients. It's the same with each little cell and what it makes. The problem of a missing or defective cell product gets bigger and more complicated when we find out that the lack of that product or any abnormal qualities in it will harm the cells that normally receive and use it.

To work right, our bodies must have foods that contain all eleven primary chemical elements, plus the trace elements needed in tiny quantities (see Table 1.2).

Table 1.2. **Chemical Elements in an Adult Body (About 70 Kilograms)**

Soft Tissue Elements	*Amount*
Oxygen	50.4 kg
Carbon	9.4 kg
Hydrogen	6.4 kg
Nitrogen	1.7 kg
Sulfur	0.3 kg
Main Bone Elements	
Calcium	1.4 kg
Phosphorus	680 gm
Magnesium	25 gm
Main Electrolytes	
Potassium	140 gm
Sodium	95 gm
Chloride	95 gm

(Continued)

Table 1.2. **Continued.**

Essential Trace Elements	*Amount*
Iron	4,500 mg
Fluorine	2,600 mg
Zinc	2,000 mg
Copper	100 mg
Iodine	25 mg
Selenium	13 mg
Manganese	12 mg
Molybdenum	9 mg
Chromium	6 mg
Cobalt	2 mg
Nonessential but Useful	
Silicon	24 mg
Vanadium	18 mg
Tin	17 mg
Nickel	10 mg
Arsenic	4 mg
Boron	4 mg
Strontium	trace
Lithium	trace
Germanium	trace

Toxic at Trace Levels

Antimony	Beryllium	Cadmium
Lead	Mercury	Thallium

Nontoxic Trace Elements in Body (Function Unknown)

Aluminum	Barium	Bismuth
Bromine	Cesium	Gold
Rubidium	Silver	Zirconium

The basic idea to remember, for your health's sake, is that we are made of the minerals and trace elements present in the soil, and unless the foods we eat are grown on rich, fully mineralized soil, our bodies will become deficient in one or more essential chemical elements, and we will become vulnerable to a disease. I don't deny that we also need exercise, fresh air, and enough rest to have healthy bodies, but our focus in this book will be on foods, the chemical elements in them, and how they are used in our bodies. If you learn about foods and the nutrients in them, the other aspects of a healthy lifestyle will tend to make more sense and fall into place in your understanding.

For example, we need to have regular elimination in order to be healthy and well, but we can't make up for a lack of chemical elements by means of bowel cleansing, such as enemas and colonics. I'm going to discuss bowel health right at the beginning because a healthy digestive and eliminative system is essential to good health.

THE CHEMICAL ELEMENTS AND COLON HEALTH

We cannot start peristaltic action, the muscular contractions that move the bowel contents along, by enemas or colon irrigation. Peristaltic activity can only be started with the consumption of foods. We must first supply foods containing fiber, chlorophyll, chloride, calcium, sodium, and magnesium to cleanse the bowel and help neutralize bowel acids. Colon irrigations without nutritional support may weaken the bowel in the course of time and may even cause injury. It is good to begin with high-sodium, high-fiber foods, which are found mostly in our fruits and vegetables, to feed the bowel wall.

Besides drinking plenty of water, remember the elements of chlorine, sodium, silicon, calcium, magnesium, and fluorine when you suffer from constipation. Otherwise, the bowel muscle tissue may develop inflammation. In fact, you may suffer bowel irregularity or even disease if you do not have enough fiber and the right chemical elements in a balanced diet, including supplements if needed.

If you must use colon irrigations, colemas, or enemas, it is good to use plenty of water mixed with flaxseed tea, and let the water flow in slowly. The water should pass along the intestinal walls. If it doesn't, it can't loosen the hardened material coating the bowel wall, and the wall coating will remain after you are through with the colon irrigation. Adequate fiber and water in the diet prevents hardening of the feces and coating of the bowel wall.

We must take constant care of the colon wall by rebuilding its tissue. This is a quality tissue restoration through proper nutrition that we are discussing. It is much better to eat more fruits, vegetables, whole grains, and legumes than to compensate for too much protein and starch by colon cleansing some years later. This is the natural way. Using the right nutrition and drinking adequate water (at least two quarts daily), you will strengthen the colon and the muscles that make up its walls. If there is a lack of chlorine in the muscles, they cannot work efficiently. Nervous frustration, which may be caused by lack of chlorine, could signal the need for more chlorine in the diet.

THE SIGNIFICANCE OF WATER IN THE DIET

Water is abundantly present in secretions, blood, serum, lymph, and all body organs, glands, and tissues. It prevents inflammation, promotes osmosis, and moistens lung surfaces for gas diffusion.

It helps to regulate body temperature, irrigates the cells and organs, and is an almost universal solvent. Water promotes all the functions of elimination. Nerves must be bathed in moisture. Without adequate water content, the blood cannot flow, waste matter won't be eliminated from the body, and many chemical processes will be disrupted. But an excess of water in the body causes pressure on and enlargement of all organs.

IMBALANCED DIETS HAVE UNDESIRABLE SIDE EFFECTS

Imbalanced diets, lacking in certain nutrients, often have consequences that are evident through symptoms and diseases that develop in the body. We may temporarily need to resort to a special diet to restore balance and get rid of unwanted symptoms. For example, when there is excessive water in the system, we may require a low-salt diet to correct it. But we must carefully limit the length of time such diets are used or the body will shift into a different but equally undesirable and imbalanced condition. It is best to work with a nutritionist or doctor who knows foods in order to take care of chemical deficiency symptoms.

WHEN LOW-SALT, DRY FOODS ARE NEEDED

When we have used beverages and salt-containing food to excess, tissues react by swelling with water (edema). Edema may result from deficiencies of protein, thiamine, and/or vitamin B_6. It also can be caused by kidney problems, congestive heart failure, pregnancy, standing too long, muscle injury, oral contraceptives, allergies, or premenstrual tension, among other causes. To diagnose the cause, your doctor first makes sure that all of the previously described medical conditions are ruled out, physical causes are checked, prescription drugs are considered for side effects, allergic reactions and premenstrual tension are discussed, and deficiencies are taken care of. The following symptoms can indicate the need for a low-salt, dry food diet:

Chills
Congestive heart disease
Cramps
Edema
Exhaustion
Fainting spells
Hemorrhoids
High blood pressure
Hypertension
Hypertrophy
Milk leg after childbirth
Muscular weakness
Nervous exhaustion
Obesity
Skin eruptions
Sleepiness
Sluggishness
Suffocating spells
Swelling under eyes
Swollen abdomen
Swollen ankles
Swollen lower limbs
Watery eczema
Weak joints
Wheezing or asthma

FOODS LOW IN WATER

A dry food diet may be helpful in taking care of edema or water-related obesity. The following foods should be used in connection with high-chlorine foods, such as celery, okra, whey, and dulce.

Acorns
Almonds
Baked rice
Barley
Beechnuts
Brazil nuts
Butter
Butternuts
Calf's foot jelly
Cardamon
Chestnuts
Currants, dry Zante
Dates
Dry gelatin dishes
Filberts

Goat's whey cheese
Grape-Nuts
Hazelnuts
Hickory nuts
Oatmeal bread
Oatmeal muffins
Olives, dried
Pecans
Piñon nuts
Pistachios
Popcorn
Puffed rice
Raisins
Raspberries, dried
Rice bran
Roasted game
Ry-Krisp
Tomatoes, dried
Veal joint jelly
Walnuts

Normally, the water content of the body is regulated by the kidneys, coordinated by a brain center that releases more or less of a certain hormone that helps regulate how much water the body keeps and how much it lets go. Edema often signals kidney or hormonal problems, which can be helped by herbs, such as KB-11, and supportive nutrition. But because edema may have many causes, you should see a doctor who understands nutrition and herbs, or a good nutritionist.

WHEN WATERY FOODS ARE NEEDED

A watery diet is helpful when we are bothered by such ailments as:

Arthritis
Constipation
Crampy tendons
Difficult urination
Dry catarrh
Dry skin
Emaciation
Lack of perspiration
Rheumatism
Rough, dry throat

Sports drinks like Gatorade are big moneymakers these days, but *Consumer Reports on Health* has reported that water is just as effective as Gatorade, unless exercise goes over an hour and a half.

FOODS HIGH IN WATER

The following are our highest water–containing foods. A diet using many such foods is needed when certain conditions prevail, as I will soon explain.

Apricots
Asparagus
Barberries
Blackberries
Blueberries
Broccoli
Brussels sprouts
Buttermilk
Cabbage, curly
Cabbage, red
Cabbage, Savoy
Carrots
Casaba
Cauliflower
Celery
Chard
Chayote
Cherries
Chervil
Collards
Eggplant
Fish
Fruit juices
Goat's milk
Gooseberries
Guava
Herbal tea
Horseradish
Huckleberries
Kefir
Kohlrabi
Leeks
Lettuce
Mandarins
Mangoes
Muskmelon
Nettles
Okra
Papaya
Parsley
Peaches
Pineapples
Prunes
Pumpkin
Radishes
Raspberries
Rhubarb
Romaine
Rutabagas
Sauerkraut
Spinach
Sorrel
Squash
Strawberries
Swiss chard
Tangerines
Tomatoes
Turnips
Vegetable juices
Watermelon
Whey

Greens are the best foods for the bowel because they contain chlorophyll, which is nature's best cleanser; beta-carotene, which is a natural cancer preventive; and fiber, which gives the bowel something to push against. When buying greens, look for fresh leaves with no tip burn, yellowing, or dark spots. Whey, either liquid or dried, also feeds the beneficial bowel flora.

Health Fact: Minnesota Is the Healthiest State

Minnesota rated number one in the nation in health care in 1999, according to Scott Morgan, president of a Kansas-based independent research and publishing company. After coming in second to Hawaii in 1998 and 1997, Minnesota broke through to the top on the basis of twenty-one health-related categories, including low infant mortality rates, low percentage of population not covered by health-care insurance, low per capita health-care expenditures, access to primary care physicians, high childhood immunization rates, and low percentage of adults who smoke. Hard on the heels of Minnesota were Hawaii, Vermont, New Hampshire, and Nebraska, in that order. The least healthy state in 1999 was Louisiana, followed by Mississippi, Alabama, Nevada, and South Carolina. Among the reasons for Louisiana's rating were the high percentage of births to teenage mothers and the high rate of sexually transmitted disease. These ratings were obtained from *Health Care State Rankings 1999,* an annual reference book that compares the fifty United States in 512 health-care categories.

CHAPTER 2

THE SOFT TISSUE BUILDERS: CARBON, HYDROGEN, OXYGEN, NITROGEN, AND SULFUR

Normally there are up to forty-six or more chemical elements that make up the human body. Five of them—carbon, hydrogen, oxygen, nitrogen, and sulfur—make up 99 percent of the body's molecules, mostly soft tissue and liquids (i.e., protein, carbohydrates, fats [lipids], and water). We will be discussing the other chemical elements in later chapters.

Whether we consider food chemistry or body chemistry, both are basically organic chemistry, which is about carbon molecules. Chemical elements, such as carbon, hydrogen, oxygen, nitrogen, and sulfur, when joined together into large molecules, leave behind their identities as individual elements to merge into the identity of their new molecular form. Before we understand what these elements accomplish together as proteins (amino acids), carbohydrates, and fats, we should first understand their individual functions.

LET'S MEET THE TEAM MEMBERS

Carbon

I know (and so do you) that carbon, by itself, exists in the form of pencil lead, charcoal, and diamonds, but this knowledge doesn't help us understand anything at all about the function of the molecules of which it is a part. Carbon in its free form is hard; yet proteins and fats are soft, and the two molecules for which carbon is best known—carbon monoxide and carbon dioxide—are both gases. Carbon is important because it can bond to four other atoms (including other carbons) in the making of molecules. Very large and complex molecules can be made that way.

Hydrogen

Hydrogen is a gas that, together with helium, makes up 99 percent of the universe. It is almost never found in its free gaseous state on planet Earth. Hydrogen and oxygen are parts of the water molecule H_2O, which covers four-fifths of the earth's surface. Hydrogen is a primary component of every acid and, together with oxygen, of every base. When we buy hydrogenated vegetable oil, we are buying a solid food spread made by adding electrically charged hydrogen atoms to liquid vegetable oil to give it a texture similar to that of butter. Hydrogen is in all proteins, carbohydrates, fats, and vitamins. It is very much a part of our body chemistry and the foods we eat.

Oxygen

This element is a gas and makes up 21 percent of the air. Oxygen also makes up from 60 to 70 percent of the human body. From the air, free oxygen is drawn into the lungs as we breathe. From there it is picked up by the iron in hemoglobin and circulated via the blood to tissues where it is used in the cells to produce energy. The waste product from cell respiration is carbon dioxide, which is carried by the blood back to the lungs, where it is exhaled. Oxygen is found in every food we eat, but it is the molecular form for building tissue, not the free form, that is carried in the blood.

Nitrogen

Nitrogen is limited, among the big three food groups, to protein as a constituent of amino acids. Carbohydrates and fats don't have any nitrogen in their makeup. But nitrogen is also found in all B-complex vitamins, choline, nitrates, and nitrites. Unlike oxygen, we don't get any nitrogen from the air, even though it makes up nearly 79 percent of it. We get our nitrogen only from foods. (Legumes, however, take nitrogen from the air and make it part of their plant protein, so indirectly we consume some of the nitrogen from the air.) Most of the nitrates and nitrites in our body come from food preservatives added to some foods. They are capable of being changed into nitrosamines, which are carcinogenic.

Sulfur

Sulfur is not found by itself in the body but is combined with other chemical elements in soft tissue protein or body fluids.

For example, sulfur is in three amino acids: methionine, cystine, and taurine (as well as cysteine, which is the oxydized form of cystine). Methionine is an essential amino acid, but cystine and taurine are not because they can be made from methionine. The absorption and availability of the essential elements zinc and selenium depend on methionine. The sulfur in methionine is also believed to slow the aging of cells. Sulfur is found in the two B vitamins biotin and thiamine, which are important in the metabolism of carbohydrates and fats. It is also found in the pancreatic hormone insulin; in the anticoagulant heparin in the liver; in the protein keratin that makes up the hair, nails, and skin; and in certain fats found in the brain, liver, and kidneys. As part of acetyl coenzyme A, sulfur assists in the energy production cycle of every cell of the body. Collagen synthesis requires sulfur amino acids. Collagen is needed to form tendons, ligaments, cartilage, skin, the linings of joints, and the protein matrices of bones and teeth. Sulfur in cystine helps protect us from radiation.

PROTEINS: THE BODY BUILDERS

Next to water, which makes up 60 to 70 percent of the body, protein is the most abundant substance at 20 percent. There are over fifty thousand different active proteins in the human body, all made out of the same building blocks—amino acids—which, in turn, are made of carbon, hydrogen, oxygen, and nitrogen, as well as sulfur, phosphorus, and iron. Some protein molecules are huge and have thousands of amino acids strung like beads on a necklace. All twenty amino acids are variations of a single basic design, an amino molecule, NH_2, combined with a carboxyl molecule, COOH (N is the symbol for nitrogen, H for hydrogen, O for oxygen, and C for carbon). The small number 2 beside the H simply means there are two hydrogen atoms. Join them together and the basic chemical formula for an amino acid is CH_3NO. The formula for the amino acid leucine, for example, is $C_6H_{13}NO_2$, a variation on the basic chemical formula. Keep in mind that of these twenty amino acids, nine must be obtained from food and eleven are manufactured in our bodies.

THE TWENTY AMINO ACIDS AND THEIR FUNCTIONS

Each amino acid has its own distinct function that works with other amino acids to build proteins. The particular combined molecular structures of the amino acids in a protein determine how it works. Of the following twenty amino acids, nine are essential (indicated by an E) in human nutrition, and the remaining eleven (indicated by an N) can be manufactured in the body.

Alanine (N)

This amino acid is an energy source for muscle tissue, is involved in sugar metabolism, and produces antibodies for the immune system. It is part of connective tissue.

Arginine (E)

Arginine assists in healing; is essential for a healthy immune system, production of growth hormone, release of insulin, and spermatogenesis; and is a precursor to the inhibitory neurotransmitter GABA (gamma-aminobutyric acid).

Aspartate (N)

Aspartate can be converted to asparagine, enhances immunoglobulin production, protects the liver, assists in DNA and RNA metabolism, and increases stamina.

Asparagine (N)

Asparagine detoxifies harmful chemicals, increases endurance, and is an important neurotransmitter. It is also a very important brain constituent, involved with the hypothalamus, which acts as a switchboard connecting sensory perceptions, emotions, intellect, memory, and other brain centers.

Cystine (N)

Cystine helps make cartilage, skin, hair, and nails; is active as an antioxidant and detoxifier; protects against radiation; and is an ingredient of glutathione, which is essential to energy production.

Glutamic Acid (N)

Glutamic acid is a neutrotransmitter that helps make up two other neurotransmitters (glutamine and GABA), reduces sugar and alcohol cravings, helps synthesize DNA, promotes healing, and helps detoxify ammonia in the brain.

Glutamine (N)

Glutamine helps make up GABA, a neurotransmitter that releases tension and brings serenity. It also assists in DNA synthesis, stabilizes blood sugar level, is a main source of energy to the small intestine, and protects against stress and anxiety. The blood contains more glutamine than any other amino acid.

Glycine (N)

Glycine makes up part of hemoglobin in red blood cells and part of cytochrome, an enzyme necessary for energy production. Along with alanine and serine, glycine stores sugar (as glycogen) in the liver and muscles. It stops sugar craving, and is a neurotransmitter. (Excess of glycine is caused by starvation.)

Histidine (E)

High in hemoglobin, histidine is a precursor to histamine (the chemical released in allergy and burns), helps maintain acid/alkaline balance in blood, and is used to treat arthritis. High histidine blood levels are associated with low zinc levels.

Isoleucine (E)

Isoleucine is required for muscle strength and stamina, is used as an energy source for muscle tissue, and is needed to produce hemoglobin.

Leucine (E)

Leucine stimulates bone healing, skin healing, and release of enkephalins (natural painkillers). It also stimulates insulin release.

Lysine (E)

Lysine helps form collagen, is essential for bone formation in children, lowers blood level of triglycerides, produces hormones, helps absorb calcium, and limits viral growth.

Methionine (E)

Methionine produces cystine and taurine; breaks down fats; reduces blood cholesterol; detoxifies the liver; is an antioxidant; and protects hair, skin, and nails. It is needed for synthesis of RNA and DNA and it assists in the breakdown of niacin, histamine, and adrenalin. It binds to heavy metals, such as lead and cadmium, and carries them out of the body.

Phenylalanine (E)

A precursor to tyrosine and thyroid hormone (thyroxine), phenylalanine acts as an antidepressant, pain reliever, and appetite suppressant; helps form collagen; and supports memory, concentration, and thinking capabilities.

Proline (N)

Proline supports the health of tendons, ligaments, and joints, and works with vitamin C to protect skin and joints. Proline also keeps the heart muscle healthy.

Serine (N)

Very important in guarding blood sugar levels, serine is needed for building and maintaining muscle tissue, produces antibodies and immunoglobulins, and is part of the makeup of nerve sheaths (myelin).

Threonine (E)

Necessary for collagen formation, threonine makes antibodies, supports the thymus gland, detoxifies and prevents fatty liver, is needed by the intestinal tract, and may be converted into the neurotransmitter glycine.

Tryptophan (E)

A precursor to the neurotransmitter serotonin, which promotes relaxation and sleep, tryptophan reduces anxiety and helps some forms of depression. It converts to niacin, lowers cholesterol, helps migraine headaches, and stimulates growth hormone.

Tyrosine (E)

A precursor to four neurotransmitters and to thyroid hormone and growth hormone, tyrosine amplifies energy, improves mental function, and may help depression caused by norepinephrine deficiency.

Valine (E)

Valine is involved with muscle function and muscle energy, fights acute physical stress, and aids assimilation of all other amino acids by the small intestine.

DISEASES AND CONDITIONS IN WHICH ONE OR MORE AMINO ACIDS IS DEFICIENT

The following diseases and conditions indicate amino acid deficiency.

AIDS
Alcoholism
Anemia
Anxiety
B_{12} deficiency
Calcium deficiency
Cancer
Candida Albicans
Chronic Fatigue Syndrome
Depression
Drug addiction
Elilepsy
Epstein-Barr virus
Excess insulin
Food allergies
Gulf War Syndrome
Herpes
High insulin level
Hypoglycemia
Hypothyroidism
Insomnia
Lou Gehrig's disease
Magnesium deficiency
Muscle spasms
Neurological deficit
Obesity
Panic disorder
Parkinson's disease
Rheumatoid arthritis
Vegan vegetarianism
Virus Infection Syndrome
Whole body radiation exposure

Children do not grow normally if they lack taurine, argenine, and cystine. Older children and adults can make cystine from the amino acid methionine.

Amino acids from what we eat are not broken down in digestion or the small bowel into smaller units. They are absorbed directly through the bowel wall into the bloodstream. Vitamin B_6 helps form some amino acids and converts others to energy.

PROTEIN DEFICIENCY SYMPTOMS: KWASHIORKOR AND MARASMUS

In the Third World, protein deficiency, known as kwashiorkor, is most common among the children of the extremely poor, especially infants who are weaned from mother's milk when another baby is born. The older child is then fed a diet of carbohydrates. These children become lethargic, apathetic, and easily irritated. The children, deprived of adequate protein, don't grow well. Sores and ulcers develop on their bodies and their bellies begin to protrude. Swelling due to edema appears in faces, arms, and legs. (Body protein normally assists in controlling fluid balance.) Growth-promoting amino acids taurine, cystine, and arginine are missing, which accounts for the lack of weight gain.

Marasmus is starvation, not just protein deprivation. After starving for six to eighteen months, marasmus victims appear far older than their years, with significant muscle wasting. This was common among American prisoners of war during World War II. With long-term starvation, sickness and disease are common, along with hair loss. Exposure to cold weather may be fatal, since no fat or protein reserves are available to produce body heat.

Adults who lack protein may feel dizzy and nauseated. Their skin becomes dry and scaly, they fatigue easily, and they become short-tempered. Infections, kidney problems, and diarrhea become increasingly worse. Since protein is the main source of sulfur intake, both kwashiorkor and marasmus produce sulfur deficiency symptoms.

Among both children and adults deficient in protein, immune system function is lowered and vulnerability to infections and diseases increases. Muscle wasting and brain damage result from long-term lack of protein, along with breakdown of the liver.

WHEN PROTEIN IS NEEDED

Protein is needed when we are too thin for our height and bone structure, or when we are depressed and lacking in get-up-and-go. Loss of vigor and stamina may be signs of deficiency. Most adults in this country know when they are not getting enough protein. It is interesting that among vegetarians, protein deficiency is rare. Combining different foods to get the right balance of amino acids requires knowledge and experience, and many vegetarians know their foods.

FOODS HIGH IN PROTEIN

The following foods are likely to supply the body with sufficient protein when used in the proper proportions. The best sources are eggs, milk and milk products, fish, poultry, and legumes. The recommended daily allowance (RDA) for protein is 56 grams for adult males and 44 grams for adult females. Always eat at least three vegetables with every protein meal.

Beans
Beef
Caviar
Cheese
Eggs
Fish
Garbanzo beans
Gelatin
Lamb
Lentils
Liver
Milk
Mutton
Nuts, raw
Peas, black-eyed
Peas, dried
Poultry
Seafood
Seeds, raw
Soybeans
Veal
Veal joint jelly
Wild game
Yogurt

Do all foods high in protein build muscle? Yes. The reason most bodybuilders focus on supplementing with arginine, glutamine, and ornithine is because they believe these three amino acids accelerate muscle building. (Ornithine is formed when an enzyme breaks down arginine.) Protein foods provide too much stimulation for some people and not enough in others. If a patient lacks protein, healing of injuries will be delayed. If a patient has too much protein, digestive problems result and there is an increased risk of obesity, kidney disease, and cancer. More is not better in the case of protein, fats, or sugar.

CARBOHYDRATES: FOODS FOR ENERGY

Low-protein, high-carbohydrate foods are important in taking care of autointoxication, kidney and liver problems, manic behavior, and low threshold anger tendencies. High-fiber carbohydrates improve bowel transit time and reduce risk of colorectal cancer.

There is no nitrogen in carbohydrates, only carbon, hydrogen, and oxygen. Hydrogen is in the ratio of 2 to 1 over oxygen in most carbohydrates. This group of foods includes table sugar, starchy foods such as potatoes and rice, and all fruits, vegetables, and legumes. Green plants use photosynthesis to combine carbon dioxide and water to form carbohydrates—molecules of varying size and complexity that provide fuel for energy, vitamins, minerals, trace elements, and fiber. Carbohydrates may be stored in the liver and muscles with the help of insulin and cortisol. When blood sugar drops below a certain point, the stored glycogen is converted back to sugar and is released into the blood. Examples of carbohydrates are glucose ($C_6H_{12}O_6$), disaccharides ($C_{12}H_{22}O_{11}$), and polysaccharides ($C_6H_{10}O_5$). Simple sugars are one or two molecules. Complex carbohydrates are made up of three hundred to one thousand molecular compounds. There are 4 calories per gram of carbohydrate.

Glucose is found in all fruits and vegetables. It is easily digested, taken up by the blood, and delivered to the cells of all organs, glands, tissues, and systems to be transformed into energy. At any given time, 25 percent of the glucose, or blood sugar, is being used by the brain. Complex carbohydrates are broken down into simple sugars by the action of enzymes. Most fruits and vegetables are classed as complex carbohydrates, as are yams and potatoes. Their main advantage as foods is that they release a slow, steady stream of sugar into the blood, which stabilizes energy production and is easy on the pancreas. They also carry fiber, which promotes bowel regularity, and many vitamins (except B_{12}), minerals, and trace elements.

Starches are great sources of glucose and include grains, legumes, and tubers (fleshy underground root vegetables like potatoes). Some doctors believe that starchy vegetables and high-sugar vegetables such as corn, carrots, peas, and beets are as much responsible for obesity as high-fat meats and dairy products. Nearly all fruits and vegetables have a little protein, but not much.

Our bodies only store a few days' worth of carbohydrate, so a high-protein, low-carbohydrate diet can be risky for some people. Raw fruits and vegetables are the best sources of vitamins and minerals. We should be eating five or more generous helpings of fruits and vegetables every day.

Low-carbohydrate, high-protein diets are popular these days, but consider the risks. Low-carbohydrate diets cause a significant (but temporary) loss of body water. When you get off the diet, the water returns. A low-carbohydrate, high-protein diet can interfere with the metabolism of calcium and can cause uremia, or blood poisoning. It can also increase ketones, which tend to make the blood acidic and encourage the deposit of fat on arterial walls, allowing ammonia to build up in the blood. When you consider dieting to lose weight, look into the risks of each diet and avoid those that compromise your health.

FOODS HIGH IN CARBOHYDRATES

High-carbohydrate foods, such as the following, are important sources of energy.

Apples, dried
Apples, fresh
Apricots
Artichokes
Asparagus
Avocado
Bananas
Barberries
Beans, kidney
Beans, string
Beets
Blackberry juice
Breadfruit
Broccoli
Brussels sprouts
Butter
Cabbage
Cardoons
Carrots
Casaba
Cauliflower
Celery juice
Chicory
Chives
Corn on the cob
Cow's milk
Cress, young
Cucumbers
Currants
Dandelion
Dates
Dewberries
Eggplant
Figs
Goat's milk
Guava
Honey, raw
Huckleberries
Kale, young
Kohlrabi
Lettuce
Loquats
Mangoes
Molasses
Mulberries
Muskmelon
Nasturtium
Nectarines
Nettles, dwarf
Okra
Olives, ripe
Oranges
Papaya
Parsley
Parsnips
Peaches
Pears

Peas, sweet
Persimmons
Pineapples
Plums
Potatoes, baked
Prunes
Pumpkins
Radishes
Rutabaga
Sauerkraut
Spinach
Squash
Strawberries
Swiss chard
Tomatoes
Turnips

CARBOHYDRATE FOODS LOW IN PROTEIN, FAT, SUGAR, AND STARCH

The following foods help reduce obesity, acidosis, fermentation, glycosuria, and other ailments caused by excess intake of fat, sugar, and protein.

Asparagus
Beans, string
Buttermilk
Cabbage
Carrots, young
Cauliflower
Celery
Chayote
Chives
Cow's whey
Cucumber
Eggplant
Fish broth
Goat's whey
Goat's whey cheese
Green juices
Kale, young
Lettuce
Liver
Milk, skim
Nettle salad
Pumpkin
Rhubarb
Rumen salad
Sauerkraut
Sorrel
Spinach
Watercress

FATS: A HIGH-POWER SOURCE

Fats, or lipids, provide 9 calories per gram as compared with 4 calories per gram for both carbohydrates and proteins. Like carbohydrates, lipids are made only of carbon, hydrogen, and oxygen. Their main purpose in the body is to act as a backup fuel source. Usually fats provide two-thirds of the body's energy. Carbohydrates contribute the rest.

Lipids don't dissolve in water, and they have to be transported in the blood by phospholipids like lecithin. They carry fat-soluble vitamins such as vitamins A, D, E, and K throughout the body.

Fatty acids have the general formula $C_nH_{2n}O_2$ and are like strings of carbon beads, each one binding to two hydrogen atoms. Each string may have as many as twenty hydrocarbon beads. A triglyceride consists of three fatty acids bound to a glycerol molecule. At this time, only linoleic and linolenic acids are recognized as essential fatty acids, which means we must get them from foods or supplements. Others may be added as scientists continue their observations and experiments to find out how individual fatty acids work in our body chemistry. There are many different fatty acids, most of them made in our bodies, but there are two categories that are very important to know about—saturated and unsaturated.

Saturated fatty acids carry as many hydrogen atoms as they can, and they harden at room temperature. Animal fats and butter are examples of saturated fats. Unsaturated fatty acids have one or more pairs of hydrogen atoms missing, and they remain liquid at room temperature. Vegetable oils, such as safflower oil, canola oil, and olive oil are unsaturated. Saturated fats and cholesterol are the main villains in cardiovascular disease.

Margarine is an unusual lipid, often made from unsaturated vegetable oil but with hydrogen atoms added to make it saturated so it will be solid at room temperature. Vegetable oil can also become solid by making unsaturated fatty acids into trans-fatty acids, which act like saturated fats.

Fatty acids help make prostaglandins, hormonelike substances that affect blood vessels, blood clotting, tissue response to hormones, and transmission of nerve impulses.

Cholesterol is another important lipid. Too much cholesterol can increase risk of heart disease. In addition to our food intake of cholesterol in meat and dairy products, our livers make as much as 2,000 milligrams of cholesterol every day. Cholesterol helps make bile, sex hormones, and the protective sheaths around nerves, but in excess it clogs major arteries.

WHAT IS YOUR FAT INTAKE?

The American Heart Association recommends that all Americans lower their fat intake to 30 percent of their total daily calories. If you eat 2,000 calories a day, your fat intake should not exceed 600 calories. The average U.S. diet is 37 percent fat, and many go up to 40 or 50 percent. Dr. Dean Ornish and the late Nathan Pritikin advocated diets under 15 percent calories from fat to prevent or reverse cardiovascular disease. I recommend no more than 20 percent fat calories per day.

Excessive fat intake leads to obesity, heart disease, and cancer. Although proteins and carbohydrates can be converted into fat after assimilation, when a person is subject to obesity, she should adopt a low-fat, low-sugar, low-starch diet.

To get a good idea of your current fat intake, you would have to keep a food journal for a couple of weeks or more. When you have a chance to examine the list of foods (and amounts) you eat over a period of time, you can get a good picture of what you can cut down and what you can cut out.

Recognize that each tablespoon of concentrated oil or fat is 90 to 100 calories. Fried foods contribute heavily to the obesity statistics in our country. If you can't keep from frying foods, get rid of your frying pan. Some cheeses are as much as 35 percent fat. Most of them are over 25 percent, so cheeses are obvious foods to cut back. I'm less concerned about fatty raw nuts and seeds because they are high in lecithin, which keeps fat in solution in the bloodstream. Brazil nuts (raw) are 63 percent oil and sesame seeds are almost exactly 50 percent oil. Black walnuts are almost 60 percent. French fries are 13 percent, larger home fries are 14 percent. Potato chips are 40 percent fat. These offer lots of possibilities for cutting back or cutting out.

I suggest using lean meats only, using fruit for dessert, and eliminating almost all cookies, pastries, and cakes; swap popcorn for potato chip snacks; have frozen yogurt instead of ice cream; avoid high-fat granola and corn bread; and use more raw vegetables and fruit.

FOODS MODERATELY HIGH IN FAT

The following foods should be eaten only in moderation:

Almonds
Avocado
Beechnuts
Brazil nuts
Butter
Cardamom
Chocolate
Coconut cream
Eggs
Filberts
Goat cheese
Goat's milk
Hazel nuts
Hickory nuts
Margarine
Olives, dried
Olives, ripe
Peanuts
Pecans
Piñon nuts
Pistachios
Popcorn
Roquefort cheese
Salmon
Swiss cheese
Walnuts

FOODS LOW IN FAT

The foods in this list reduce obesity, acidosis, ketonuria, fatty tumors, fatty anemia, rheumatism caused by fatty acids, and other health problems caused by an excess of fat in the body.

These foods are:

Apples
Apricots
Artichokes
Asparagus
Baked potatoes
Barberries
Bass
Beans, string
Beets, cooked
Blackberries
Blueberry juice
Breadfruit
Brown cheese
Brussels sprouts
Cabbage
Cabbage, celery
Cabbage, curly
Cabbage, red
Cabbage, Savoy
Calf's foot jelly
Cardoons
Carrots, young
Cauliflower
Celery
Celery hearts
Celery juice
Chard
Chayote
Cherries, light
Cherry juice
Chervil
Chicory
Chives
Collards
Corn
Cow buttermilk
Cress
Cucumbers
Currants
Custard apple
Dandelion
Dewberries
Dill
Egg whites
Eggplant
Figs, black
Fish, lean
Gelatin
Goat buttermilk
Goat cheese
Goat cottage cheese
Goat's milk, skimmed
Green turtle
Greens
Huckleberries
Kale
Kidney beans
Kohlrabi
Lamb, lean
Lamb's lettuce
Lettuce
Loquats
Lutefisk
Mangoes
Mangosteen
Marjoram
Mirabelles
Mulberries
Nasturtium
Nectarines
Nettles, dwarf
Okra
Onions, steamed
Papayas
Parsley
Peaches
Pears
Peas, green
Persimmons
Pineapple
Plums
Pomegranates

Prunes
Quince
Radishes, young
Red snapper
Rice bran
Romaine
Rutabaga
Salsify
Sapotes
Sauerkraut
Shallots
Sorrel
Spinach
Strawberries
Turnips
Veal jelly
Watermelon
Whey
Wintergreen
Yeast

FOODS LOW IN SUGAR AND STARCH

For many reasons, but mainly because of adult-onset diabetes and hypoglycemia, we should be careful about our sugar intake. Americans averaged 26 pounds of sugar per year in 1900. In 1998, the average person used six times that amount. The Center for Science in the Public Interest estimates that sugar amounts to 16 percent of the average American's dietary calories and 20 percent of the average teenager's calories. The foods that are lowest in sugar and starch are as follows:

Beans, string
Beef
Beef broth
Bone broth
Butter
Calf's foot jelly
Cheese
Cucumbers
Egg whites
Fish
Fish broth
Fowl
Gelatin
Goat meat
Hickory nuts
Lamb
Lettuce
Limes
Liver
Loquats
Lutefisk
Mullet
Mutton
Nasturtium
Nettles, dwarf
Pork
Spinach
Tomatoes
Wild game

FOODS HIGH IN PROTEIN AND COMPLEX CARBOHYDRATES

Sometimes it may be necessary to exclude from our diet foods that are high in fat and sugar, the latter including carbohydrates that rapidly break down complex sugars into the basic sugar glucose for assimilation into the blood. When dietary sugar is assimilated too rapidly, excess insulin is released from the pancreas, which causes three undesirable effects. (Insulin in normal amounts assists cellular assimilation of glucose and stores the excess for future use.) Excessive insulin causes at least some of the surplus blood sugar to be converted into fat for storage, blocks stored fat from being burned up, and stimulates an increase in the liver's production of cholesterol. These undesirable effects increase the risk of obesity, diabetes, heart disease, and other health problems. The following list of foods can be of great importance in preventing the release of excessive insulin.

Asparagus
Beans, string
Bone broth
Cabbage
Calf's foot jelly
Cauliflower
Celery
Chayote
Chicken bone broth
Chicken, skinless
Cucumbers
Distilled water
Egg whites
Eggshell broth
Endive
Fish
Fish broth
Gelatin
Greens juice
Herbal teas
Kale
Lettuce
Nasturium
Nettle salad
Radishes
Red meat, lean
Sauerkraut
Sorrel
Spinach
Turtle broth
Watercress

When we become overweight, part of the problem may be that we are combining sugary foods with fatty foods, including fatty proteins. This accelerates fat storage. We need to eliminate refined carbohydrates such as sugar, white flour products, and white rice from the diet. Dr. M. C. Bethea, Dr. L. S. Balart, and Dr. Sam Andrews believe that even complex carbohydrates like potatoes, yams, beets, carrots, peas, and others that are high in natural sugars should be screened from the diet. These doctors, together with H. Leighton Steward, present their own diet plan in *Sugar Busters!: Cut Sugar to Trim Fat.* In my own experience, using too many sugar-containing or high-starch foods together with fatty foods often results in sleepiness, lethargy, fatigue, weakness, and loss of motivation to exercise or do physical work. That is when a diet limiting or eliminating sugary, starchy, or fatty foods and emphasizing high-protein foods and low-starch complex carbohydrates can be of help. We get enough fats and oils from vegetables, legumes, whole grains, raw nuts and seeds, and lean meat to fulfill our bodies' needs.

Food Facts

In a recent year, Americans put $5 billion in coins into 10 million dispensing machines to purchase 10 billion servings of soft drinks. That represents from 50 to 100 billion teaspoons of sugar. An article in *Food Technology* magazine estimated that 40 percent of American one- and two-year-old children drink half a 12-ounce can of soda pop per day, which would contain from 2 to 5 teaspoons of sugar and 25 to 50 milligrams of caffeine, which is suspected of being capable of affecting the brain in developing children. In 1900, few Americans drank more than a glass or two of lemonade over a year's time and perhaps a glass of real root beer on a rare occasion. In our time, nearly three-quarters of the sugar we use has already been added to the products we buy. The average American consumes about 151 pounds of sugar yearly.

CHAPTER 3

THE ELECTROLYTE TEAM: POTASSIUM, SODIUM, CHLORINE, AND OTHERS

Potassium, sodium, and chlorine are the three dominant electrolytes in the human body. Other electrolytes are magnesium, bicarbonate, phosphate, sulfate, and proteins. Electrolytes are electrically charged atoms or molecules that conduct electricity as they dissolve in liquids such as water or blood. A Swedish scientist named Svante Arrhenius won a Nobel prize for his research and discoveries about electrolytes. He called the electrically charged particles "ions"—positively charged particles were "cations," and negatively charged particles were "anions."

There are many electrolytes dissolved in the blood, lymph, plasma, tissue fluids, and cellular fluids, but we will only deal with those involved with cell metabolism in this chapter. (Technically, all electrically charged ions in body fluids are electrolytes, but not all of them are involved with cell metabolism.)

Seldom do we find that one of the three main electrolytes is deficient in a healthy body, and all of them are relatively common in the foods most of us eat in this country.

If you check any food composition table, you would see that potassium and sodium are found together in every food category. Both in foods and in the human body, they are often accompanied by chloride, which is the chemically active form of the element chlorine. All the natural foods I can think of have a lot more potassium than sodium, but they all have both—and chloride. Processed foods (potato chips, breakfast cereals, roasted nuts, soft drinks, etc.) are the only foods that have more sodium than potassium, and I believe you can guess why. The manufacturers often add salt to their products. Why? Table salt (sodium chloride) acts as a preservative and a flavor enhancer. Products have a longer shelf life, and, besides, the salty taste is popular and helps sell many commercial food products. Salt is, however, somewhat addictive, and excessive intakes of it complicate body chemistry and increase the risk of high blood pressure in some individuals.

Using highly salted foods forces the body to get rid of excess sodium chloride, usually through the kidneys, in order to keep the proper overall balance of electrolytes, the right acid/alkaline balance, and the right amount of water in the body. The problem is that when sodium is excreted, it doesn't depart all by itself.

Other chemical elements like potassium, chloride, and phosphorus are excreted, too, not to mention the water that must be used to carry them out of the body. By the time the amount of sodium returns to normal in the body, there may be a temporary deficiency of the other elements excreted along with it and possibly a degree of dehydration, resulting in further imbalances. The body's constant activity of working toward overall balance in all its systems is called *homeostasis.*

WHAT ELECTROLYTES DO FOR YOU

All body cells, even the bone cells, are mostly fluid inside and are surrounded by fluid on the outside. There is a lot of potassium and a little bit of sodium and chloride inside every cell in your body, and, similarly, there is a lot of sodium and chloride and a little bit of potassium in the surrounding fluid outside each cell. There is a reason for this that we will discuss later, but for now I want you to remember that every cell of your body has sodium, potassium, and chloride in and around it. In a 70-kilogram person (154 pounds), there are normally about 140 grams of potassium, 95 grams of sodium, and 95 grams of chloride distributed throughout the body.

Sodium, potassium, and chloride work together in all body tissues and fluids, with a kind of dynamic tension among the three. As I have pointed out, sodium and chloride are highly concentrated in the fluid outside the cells of the body while potassium is highly concentrated in the fluid inside them. There are other electrolytes involved also, but the same concentration of electrolytes is maintained in the fluid inside and outside the cells. Electrolytes interact with cell membranes to allow nutrients to enter cells and wastes to leave. When a concentration imbalance occurs, water either enters or leaves the cells until electrolyte concentrations are the same inside and outside the cells. (See Table 3.1.)

Nerve cells, unlike other cells, are long and narrow, with electrolytes concentrated at both ends of the cell but not along the middle. We speak of nerve cells being *polarized* instead of electrolytes being uniformly distributed around the inside and outside of nerve cell membranes.

Inside the end of each nerve cell are the potassium and other electrolytes, while outside are the sodium and chloride, creating an electric potential at the gap (synapse) that a nerve impulse has to "jump over" as it travels from one nerve cell to the next. When that electric nerve impulse arrives, sodium ions (electrically charged sodium atoms) rush into the nerve cell, potassium ions rush out of the nerve cell into the synapse, and the cell membrane is depolarized. The depolarization allows the nerve impulse to jump across the synapse and enter the next nerve cell, usually with the help of a chemical called a neurotransmitter. When the nerve impulse has crossed the gap between nerve cells (synapse), the potassium ions rush back into the cell, the sodium ions rush out into the synapse (outside the cell), and the normal polarization of the nerve cell is restored. It is again ready to respond to the next nerve impulse that arrives.

Table 3.1. **Electrolytes Outside and Inside Cells**

Electrolyte	*Outside Cell (Milligrams per Liter)*	*Inside Cell (Milligrams per Liter)*
Sodium ($Na+$)	3,266	230
Potassium ($K+$)	195	5,772
Calcium ($Ca++$)	100	40
Magnesium ($Mg++$)	36	40
Chloride ($Cl-$)	3,646	trace
Bicarbonate (HCO_{3-})	1,647	488
Phosphate (PO_{4-})	384	6,528
Sulfate (SO_{4-})	48	trace

Below our conscious awareness, billions of electrochemical processes take place in our bodies every second, and if we are eating right and pursuing a healthy lifestyle, chances are these microevents are helping us feel good about our lives.

THE HAZARDS OF ELECTROLYTE IMBALANCE

We can't really take the chemical balance of our bodies for granted. Many people have died from loss of electrolytes and dehydration due to prolonged diarrhea, vomiting, side effects of some drugs, or the impact of a disease. The popular singer Karen Carpenter died of cardiac arrhythmia in 1983 due to electrolyte imbalance brought on by chronic anorexia, bringing public attention to the danger of this affliction.

Infants are especially vulnerable to dehydration and electrolyte loss due to diarrhea. In a recent year, three million children worldwide, most from Third World countries, died from diarrhea due to various causes. Diarrhea in infants may be due to cholera, parasites, gastritis, a type of *E. coli* bacteria, rotavirus, and other viruses. In the United States, fifty-five thousand children were hospitalized with rotavirus in 1997, and forty of them died.

In this country, parents sometimes fail to take lengthy bouts of diarrhea seriously enough to consult their doctor. Fluid loss in an infant's watery stool causes dehydration and electrolyte loss. When fluid loss in an infant reaches 15 percent, the heart collapses and the baby dies. Most parents believe diarrhea is not particularly dangerous, and some have waited too long before contacting a physician. What signs should people look for to decide whether they should call a doctor?

Your child's pediatrician should be called if the baby has diarrhea more than once every two hours for twelve hours, or has a temperature over 102.5 degrees along with the diarrhea for twenty-four hours, or refuses to eat or drink. If there is blood in the stool, or the baby shows signs of dehydration or cries in pain, call the doctor. Even if the child only has mild diarrhea, phone your doctor if it goes on for two weeks or more. The medical experts say to keep feeding the baby every day while the diarrhea is continuing. Give him a little water to drink after each watery stool. Never give your baby any medication that you think would help stop the diarrhea. If you feel like you have to do *something,* then contact your pediatrician.

In people of all ages, potassium depletion below a certain level disrupts heart function, and death soon follows. Loss of other important electrolytes, magnesium, and calcium, may also contribute to heart arrhythmia and death. Dehydration tends to force a greater loss of potassium than of sodium, which raises the sodium level in all tissues, but most importantly in the heart muscle where it increases the risk of arrhythmia and death.

When there is too much sodium in a person's body, the kidneys get rid of the excess, but they always have to expel some potassium, chloride, and water along with the sodium, even if it means leaving the body deficient in potassium. (No chloride deficiency has ever been recorded.) After you have a look at the following food table, I will discuss the electrolytes individually, with the intention of giving you additional reasons to be more intentional and careful in your food shopping, preparation, and consumption.

FOODS HIGH IN SODIUM, POTASSIUM, AND CHLORIDE

- Blackberry juice
- Bone joint jelly
- Caraway
- Cheese
- Chicken broth
- Currants, dry Zante
- Dandelion
- Eggshell broth
- Goat cottage cheese
- Herring
- Roquefort cheese
- Rye bran muffins
- Salmon
- Sauerkraut
- Swiss cheese
- Tomatoes, dried
- Veal jelly
- Veal joint/whole rice jelly
- Veal shank
- Whey powder
- Wild cherry juice

POTASSIUM: THE ALKALIZER

Potassium makes up 5 percent of the mineral content of the body, and we need to get 2.5 grams of it in our daily diet. I call it the alkalizing element because it does the most work of any element in maintaining the acid/alkaline balance in the body. It is one of our most important electrolytes (the others are sodium, chloride, calcium, and magnesium) in the body, functioning inside cells to help nutrients pass in through cell walls and to help wastes pass out. It participates in the passage of nerve impulses from one nerve cell to another, works with sodium to help maintain the right acid/alkaline balance in the body, and helps maintain the correct water balance. It teams up with phosphorus to get oxygen to the brain and works with calcium in the muscles to regulate nerve-muscle interaction. Potassium contributes to maintaining healthy skin. The average person in the United States may not get sufficient potassium in her daily diet.

Normal growth requires potassium involvement in enzyme activities. It plays a part in making muscle protein from amino acids, assists in the storage of glucose in the liver, and cooperates with sodium in maintaining blood pressure. It helps in the synthesis of nucleic acids and signals the kidneys to eliminate wastes in the urine. Potassium works with sodium to regulate the heartbeat.

Since the kidneys regulate levels of sodium and potassium in the body, there are only a few ways that a toxic excess of potassium can develop in the body.

An excess of potassium can develop due to kidney failure, extreme dehydration, infection, adrenal insufficiency, and bleeding in the bowel. A dysfunction in protein breakdown can cause excess potassium, as can insulin under certain conditions.

Depletion of potassium can take place when excess sodium intake crowds it out of the body, by prolonged diarrhea or vomiting, and by excessive sweating or use of diuretics. Diuretics are herbs or drugs that increase urine excretion. Alcohol, coffee, and sugar are antagonists of potassium. Alcoholic beverages and coffee are diuretics, which cause the body to excrete more water than it takes in, causing dehydration and loss of potassium. Also, when alcohol intake depletes magnesium, then a certain amount of potassium is also lost along with the magnesium.

Low blood sugar causes the adrenal glands to react, which, in turn, causes more potassium to be lost. Hormones, such as aldosterone and cortisone, can cause deficiency. Some doctors believe potassium deficiency is relatively common in the United States, and believe this deficiency is dangerous.

DEFICIENCY SYMPTOMS

If potassium becomes too deficient, changes in body chemistry, water balance, and possibly blood pressure will take place. The sodium content of the heart and other muscle tissue will increase. The heart rate will diminish, and there will be generalized weakness. Reflexes are poor, brain function is impaired, and muscles become soft and unresponsive. Sterility and kidney problems appear. If extreme dehydration is involved, potassium can become drained and proteins break down. The risk of stroke is increased. As I have pointed out before, extreme potassium deficiency can be fatal. If ever you have reason to believe you are low or deficient in potassium, Table 3.2 should be helpful to you.

SODIUM: THE YOUTH ELEMENT

I call sodium the "youth element" because it helps keep the joints limber. Sixty percent of the body's sodium is in the fluid that surrounds our cells, 10 percent is inside our cells, and the

Table 3.2. **Ten Foods Highest in Potassium**

Food	*Milligrams per 100 grams*	*Food*	*Milligrams per 100 grams*
Dulse	8,060	Wheat bran	1,121
Kelp	5,273	Sunflower seeds	920
Goat's whey	3,403	Wheat germ	827
Black strap molasses	2,927	Almonds	773
		Raisins	763
Rice bran	1,495		

other 30 percent is in the bones. We need 2 grams of this element daily. Sodium from foods or table salt is almost totally assimilated into the body. Unfortunately, most people use from 5 to 15 grams of sodium daily, and that's too much. This excess is not exactly toxic, but it concerns most doctors and nutritionists, including me.

Normally, an adrenal hormone called aldosterone signals the kidneys when to conserve sodium, so the kidneys eliminate excess sodium in the urine until they get the signal to stop. The problem is that a certain amount of potassium is always eliminated with the excess sodium, which creates the danger of potassium deficiency. Along the same line, too much sodium intake increases the risk of hypertension in individuals whose body systems react to sodium above a certain threshold. (Not everyone who uses too much salt acquires hypertension.)

While chronic sodium deficiency is rare, short-term deficiency is not uncommon. Working at hard manual labor for eight hours on a hot day over 100 degrees can cause a loss of 8 grams of sodium in perspiration. Professional athletes experience the same rate of sodium loss during hotly contested games, which is partly why TV sports fans see them drinking Gatorade, a commercial, high-carbohydrate drink with 440 milligrams of sodium and 120 milligrams of potassium per liter. Starvation, vomiting, and diarrhea can also cause short-term sodium deficiency.

I feel it is very important to know the difference between the way the body responds to table salt and the way it responds to sodium in the foods we eat. Foods usually deliver vitamins and minerals to us in a mixture of mutally enhancing nutrients, never just one nutrient. In spinach, for example, sodium is accompanied by vitamins A, B-complex, C, and K, and minerals copper, calcium, iron, magnesium, phosphate, zinc, and potassium, some amino acids, carbohydrates, and fatty acids. These nutrients interact as they are assimilated, enter the bloodstream, pass through the liver, and go on to their cellular targets. Table salt, on the other hand, contains sodium chloride (the chemically active form of chlorine), calcium silicate (to keep salt from caking), and a trace of potassium iodide (to prevent goiter). Table salt enters the body more like a drug, and the side effects are water retention, adrenal gland stimulation, kidney stimulation, and elimination of potassium, to mention a few. I believe in food sodium and appreciate its benefits, but I do not approve of table salt use. Each teaspoon of table salt (about 5 grams) contains 2 grams of sodium.

Food sodium helps keep calcium in solution, preventing it from depositing in joints or developing spurs. Sodium is required in the metabolism of proteins and carbohydrates, and a great amount of sodium is used in bone building. Electrically charged sodium ions help escort amino acids through cell membranes for use in making proteins. (Remember, our cells manufacture fifty thousand or more different proteins.) Sodium lines the stomach and bowel walls, protecting them from acid damage. Sodium is part of every body secretion—saliva, tears, perspiration, mucus, catarrh, and sexual fluid. It is essential for the spleen and liver, and helps maintain osmotic pressure. Sodium works best in an environment of other nutrients that accompany it in foods. In its purified form as table salt, I feel sodium chloride, taken in excess, does more harm than good.

I encourage you to use seasonings on your foods other than table salt. "Lite" salt, either made with all potassium chloride or a mix of potassium chloride and sodium chloride, is an improvement over table salt, although not a very good improvement, in my opinion. There are herbal salt substitutes such as Vegit, Mrs. Dash, and many others at local health food stores. Look over what is available and buy a few now and then to find out which ones please your palate. You'll be much healthier if you lower your total salt intake to 2 grams daily.

SODIUM DEFICIENCY

Lack of sufficient food sodium may result in joint stiffness, rheumatism, neuralgia, and bladder ailments. The hydrochloric acid in the stomach can't be made without sodium chloride. Sodium acts on the nerves, secretions, membranes, stomach, intestines, and pancreas.

Sodium deficiency symptoms include weakness, gas, nausea, vomiting, heart arrhythmias, attention deficit, poor memory, and difficulty in concentration. If this occurs in conjunction with dehydration, water should be given before the sodium deficit is taken care of.

If you should ever need to increase your intake of food sodium, Table 3.3 lists some foods to consider.

Table 3.3. **Ten Foods Highest in Sodium**

Food	*Milligrams per 100 grams*	*Food*	*Milligrams per 100 grams*
Kelp	3,007	Goat's whey	371
Green olives	2,400	Scallops	265
Ripe olives	828	Cottage cheese	229
Sauerkraut	747	Lobster	210
Cheddar cheese	700	Swiss chard	147

CHLORINE: THE CLEANSER

It seems odd that chlorine is an irritating, acrid, deadly gas, while its chemically active form, chloride, is part of the most widely used seasoning in the world—salt. I call chlorine "the cleanser." One of its important functions in the stomach is to destroy harmful microorganisms (that accompany food) before they can get a foothold in the body. Salt (sodium chloride) is 60 percent chloride. About 20 percent of the chloride in our bodies is in organic compounds.

Together with sodium, chloride aids in stabilizing the pH of the blood at 7.35 to 7.45, just slightly alkaline. Hydrogen, together with chloride, forms the hydrochloric acid in the gastric juices, which breaks down proteins, collagen, and sucrose. Chloride promotes normal heart function.

I am not going to recommend foods high in chloride because we always get more than enough chloride when we eat foods that contain sodium and potassium. An overdose of chloride can cause weakness, confusion, and coma, but excess chloride is usually eliminated in the urine.

CALCIUM: THE KNITTER

Calcium is primarily a bone builder, but it is also in and around every cell of the body (mostly in tissue fluid outside cells) as an electrolyte. Next to sodium, potassium, and chloride, calcium, and magnesium are the most important electrolytes. In body fluids, both of these elements carry twice the positive electrical charge carried by sodium or potassium. As a salt, calcium is found in lime (calcium oxide), chalk (calcium carbonate), and bone material (calcium phosphate). The parathyroid glands control blood levels of calcium. This particular element is very difficult to assimilate from either foods or supplements. Vitamin D must be present for calcium to be assimilated, but copper, manganese, and zinc are also involved in calcium intake.

Only a small amount of food calcium or calcium supplement is actually taken in, and the most recent research shows that most Americans are deficient in calcium and should be taking 2 grams daily or more in foods and supplements. More calcium is assimilated when it doesn't have to compete with other minerals being digested and taken in. This has prompted many nutritionists to suggest that we take calcium supplements with a between-meal snack of yogurt, milk, or orange juice and not at main meals or along with any multivitamin/mineral supplement.

The best calcium foods are milk products. All other food sources are relatively low in calcium content. Salmon and sardines (with cooked bones) are fair sources. Broccoli, kale, turnip greens, almonds, and figs are among the highest plant food sources of calcium.

MAGNESIUM: THE RELAXER

The same magnesium that causes flares to burn with such a brilliant white light at night also flourishes in the body as an important electrolyte, nearly always with or near calcium and phosphate, forming the same trio that makes up bone.

Magnesium is almost equally present inside and outside the cells of our bodies. Actually, every element in the body, excepting perhaps the cobalt in vitamin B_{12}, has multiple tasks and relationships to work out, and magnesium is no exception. I have dubbed magnesium "the relaxing element" because of its calming influence. Some of you readers may have recognized magnesium sulfate as Epsom salts, dissolved in hot water and used as a mineral bath to relax in. Magnesium, as an electrolyte, is important for normal heart function and in enzymes required for metabolism of proteins, carbohydrates, and fats. Magnesium-deficient diets have caused significant heart and artery problems in experiments with lab animals, and the same deficiency in humans is suspected of producing cardiovascular disease that is so rampant and so deadly in our country.

The question is, how could so many Americans possibly be deficient in magnesium since it is so prominent in leafy green vegetables, legumes, whole grains, raw nuts, raw seeds, and seafood? The answer is simple: Most Americans don't eat much of the foods I just named. I am personally impressed with studies that show people with adequate-to-generous intakes of magnesium have fewer heart attacks. I have advocated using more fresh, raw fruits and vegetables and less "meat and potatoes" for many years, and I believe (although I haven't polled my students and ex-patients) that those who follow my diet advice have far less cardiovascular disease than those who don't.

BICARBONATE

Bicarbonate is inside our cells in small amounts and is in the fluid outside cells in large amounts. It acts as a buffer in the lymph and blood to maintain acid/alkaline balance and takes part in the follow-up activity to cellular respiration. In cellular respiration, oxygen is combined with carbohydrate in a series of steps that result in the production of carbon dioxide, water, and energy. The carbon dioxide waste is transformed into bicarbonate or carbonic acid so it can be carried by the blood to the lungs, transformed back to carbon dioxide, and exhaled. We all have plenty of bicarbonate electrolytes from foods and from the metabolism of carbon dioxide in the body.

PHOSPHATES

Phosphates are the most numerous electrolytes inside the cells of the body, possibly because adenosine triphosphate (ATP) is so important in the energy production of cells. Each cell maintains only a small reserve of ATP, but it has an enormous number of phosphate molecules with which it can make more ATP when it is needed. Phosphates are able to store the energy produced by cell respiration. B-complex vitamins and many enzymes function only in the presence of phosphates.

SULFATES

There is a minute number of sulfates in the fluid surrounding cells. They are possibly breakdown products from sulfur-containing insulin or amino acids, but they still help maintain osmotic pressure, blood pressure, and water balance as all electrolytes do.

OTHER ELECTROLYTES

There are a number of other electrolytes in the form of electrically charged organic molecules inside and outside of cells. When we consider that there are about four thousand enzymes in each cell obeying the blueprint for cell function laid out in the DNA, making energy and proteins under the direction of the RNA, there is a lot of work requiring transport of materials back and forth through cell membranes. Many of these electrolytes are involved in transporting nutrients into the cell, wastes outside the cell, and products designated for other tissues through the bloodstream to the target cells.

LIFE IS AN ELECTROCHEMICAL PROCESS

Everything important that happens in the body is directly related to electrons being moved from place to place to make things happen. Electron transfer is the basis for all oxidation and reduction reactions, including cellular respiration—the "making" of energy. Electrolytes are the top enzyme activators and are essential to the movement of nerve messages in the body and protection of the heart muscle and blood vessels.

CHAPTER 4

THE BONE MAKERS: CALCIUM, MAGNESIUM, PHOSPHORUS, AND OTHERS

In my work as a chiropractor, before I actively emphasized nutrition with my patients, I sometimes encountered patients who couldn't hold a spinal adjustment. Their vertebrae would slip out of alignment a day or less after they had come to my office. I was mystified about the cause, especially since I only ran into it every now and then. I would adjust that patient a second, third, and fourth time, and he would telephone me each time to tell me "it didn't take." The majority of the patients who came to me with back problems were holding their adjustments much longer. I kept wondering what I could do to help those whose vertebrae were not remaining in place. Finally, the nutritional studies I'd done with Dr. Rocine gave me an idea.

I put all of my difficult patients on a high calcium diet—goat's milk, yogurt, cheese, nut and seed butters, kale, broccoli, and cod liver oil to help assimilate the calcium. Soon I was seeing much improved results. The spinal adjustments began to hold as long as they did with my other patients. I can tell you, I had some very happy patients!

From that time on, I routinely advise every new patient who can't hold an adjustment to include more high-calcium foods in her food regimen. Chiropractic adjustments, osteopathic work, massage, acupuncture, homeopathy, herbal support, surgery, physical therapy, and all of the other healing arts have their rightful place, but no healing can be successful without adequate nutritional support for patients. Only foods build, strengthen, and repair damaged tissue.

In my experience, most people believe that the purpose of our bones is to provide a convenient structure to hold the rest of our body parts in place so we can dance, run, jump, sit, and, in general, do the things we have to do and the things we like to do. Bones are just "there," and we don't pay much attention to them, unless we break one or throw our backs out of alignment. I also think most people would be very surprised to find out that bones are just as much alive as muscle tissue, that they are made up of living cells like other parts of the body, and that, despite the old saw about being "dry as a bone," bones are nearly half water. All blood cells, red and white, are produced in the marrow of our bones. Bones are just as important as our hearts or lungs, and they play a vital role in supporting our immune systems and body chemistry! In this chapter, we're going to find out how every cell in our bodies depends upon our bones to keep them alive, and how hormonal messengers are sent to the bones when our cells need help from them.

Calcium, phosphorus, and magnesium are found in the body and bones in significant amounts, so we will call them macrominerals. There are about 1.4 kilograms of calcium, 680 milligrams of phosphorus, and 25 milligrams of magnesium. In contrast to the macrominerals, there are many trace elements deposited on our bones.

Bones are actually living protein networks to which minerals attach themselves. Not all of the minerals deposited on bones are essential to bone building. There are at least two dozen elements in bones that have no known function in the human body, as well as a handful of nonessential elements, such as boron, strontium, silicon, barium, bismuth, and arsenic (yes, arsenic), that are believed to do some good. Five toxic elements—lead, cadmium, mercury, polonium, and radium—are often found in human bones. As long as they are stabilized in the bones, they do no apparent harm.

All in all, there are at least forty-one elements in most human bodies, of which only twenty-one are known to be essential. We cannot assume that just because an element is found in the human body, it must serve some purpose. Our natural defenses, such as the white blood cells, antibodies, antioxidants, and macrophages, don't attack or immobilize toxic or nonessential chemical elements and remove them from the body as they do with harmful bacteria or viruses. We pick up some of these unneeded or unwanted elements from our food, water, and air, and they simply settle in the body, usually in the fatty tissue, liver, or bones. There are ways of getting them out, which I will mention later.

Bones not only provide structure to our bodies, but they act as warehouses holding reserves of essential chemical elements. Any excess of essential chemical elements in the blood is either excreted in the urine or deposited on the bones. When there is a shortage of bone minerals in the body, special hormones signal for the right elements to be released from storage in the bones to be used where needed.

Nearly 99 percent of the body's calcium, 80 percent of its phosphorus, and over 50 percent of magnesium is in the bones or teeth (in the form of calcium carbonate, calcium phosphate, and magnesium phosphate). Thirty percent of the body's total sodium is also stored in the bones. These elements can literally be transferred out of the bone "warehouse" at a moment's notice and shipped to cells that need them.

Because calcium, phosphorus, and magnesium are essential to so many of the body's life-sustaining chemical processes, it is of crucial importance to have enough of all three to meet the body's needs at all times. What happens when we become deficient in one or more of these essential elements? Let's have a look.

Deficiency of an essential bone element anywhere in the body triggers a reaction in which one or more hormones signal the release of that element from bone to meet the need. If the intake of one or more of the bone elements has been chronically low, the continual "borrowing" of that element to meet body needs will weaken the bone. The body has a system of self-protective priorities. It will seemingly "rob" nutrients from a lower-priority part of the body to sustain a higher-priority part's need, even to the harm of that lower part. The highest-priority part of our anatomy is the central nervous system (brain and spinal cord). It would be the last part of the anatomy to suffer depletion in case of long-term starvation.

To offer a specific example involving the bone structure, osteoporosis is caused by excessive "borrowing" of calcium from the bones. The ultimate cause is not always and not only insufficient calcium in the diet but interference with the various factors that need to be in place before adequate calcium can be assimilated from nutrients in the small intestine. Celiac disease (sprue) blocks calcium and other minerals even when they are abundant in the diet.

All three main bone elements are locked into a close reliance on and interaction with one another, in which any change in the amount of calcium, phosphorus, or magnesium affects the other two. In a healthy body, calcium and phosphorus remain in a 2 to 1 ratio, but if lack of vitamin D prevents food calcium from being assimilated into the body, the balance is upset and hormones are released in an attempt to restore balance. If magnesium isn't present, vitamin D (a hormone) alone can't assist calcium assimilation. Excessive use of antacids depletes phosphorus, and again the balance is disrupted. Phosphorus is essential to energy production by cells, but this energy production can only take place with the help of magnesium-activated enzymes. When any of these three elements gets too low in the fluid inside or outside the cells of the body, the kidneys conserve blood levels, and assimilation of the needed element from the bowel increases. There are medications and diseases that interfere with any or all of the big three bone builders. Any shortage—even a slight shortage for a brief period of time—of an essential element will do some damage in the body.

I don't think it is possible, except in extreme cases of starvation, for any mineral, trace element, or vitamin to be completely used up and no longer present in the body. That's not what is meant by deficiency. The actual condition of deficiency develops when there is too little of a nutrient to meet all the body's different needs for it, and when that happens, some body functions are impaired more than others. Now, let's have a closer look at each of the main "bone-maker" elements and their less publicized (but still important) helpers.

CALCIUM: THE KNITTER

I call calcium "the knitter" because of its influence on healing, not only of broken bones but of wounds, lesions, and disease damage in the body. Besides building bones, calcium is a very important electrolyte, which is needed by several enzymes and hormones. As an electrolyte, it is involved with nerve transmission, water balance, acid/alkaline balance, and maintaining osmotic pressure. Calcium is necessary for blood to clot and for heart muscle function. It may help maintain blood pressure and reduce the risk of colon cancer. It is not well assimilated from the bowel, a fact that points to a larger problem—the difficulty of getting and keeping enough calcium from a standard diet to meet all body needs.

The discovery in the not-too-distant past that 20 percent of all postmenopausal women have acquired osteoporosis (porous bones weakened by severe calcium withdrawal) has stirred government agencies and private researchers to look into how much calcium people are getting in their daily diets and how realistic our reference daily intake (RDI) of 1,000 milligrams is for today's health needs. The National Institutes of Health found that the average female was getting 635 milligrams of calcium, 365 milligrams below the RDI, and recommended increasing the RDI to 1,500 milligrams daily. However, getting people to eat more high-calcium foods or take calcium supplements is not something the government can mandate. Nor is calcium assimilation just a simple matter of intake.

For example, we seem to need at least some minimum of exercise to "coax" calcium into the body. It has been demonstrated that calcium is assimilated better by people who exercise regularly. (NASA found that out the hard way when their astronauts returned from orbiting the earth with significant calcium deficiencies despite an adequate daily calcium intake during space flights.) Estrogen helps women assimilate more calcium, and testosterone produces the same result in men. Estrogen and testosterone therapies, respectively, have increased bone mass in women and men diagnosed with osteoporosis. Men tend to get osteoporosis later in life than women because men tend to build larger bones in their younger years than women do. Larger bones simply tolerate calcium loss longer than the more slender bones of most women. Older men, however, account for 30 percent of broken hips reported in medical statistics. Broken hips in both sexes make up one of the significant markers for osteoporosis.

Osteomalacia—the adult form of rickets—is another calcium deficiency disease. Instead of making bones more brittle and porous as osteoporosis does, osteomalacia makes bones more flexible, resulting in deformities and pain. The cause is usually vitamin D deficiency, and the most effective reversal is short-term use of high dosages of vitamin D, gradually reducing dosages to 400 IU (International Units) daily. This should always be done under the supervision of a doctor since vitamin D is a very potent hormone and can become toxic if improperly used.

A double-blind experiment at the University of California at San Diego, using a daily supplement of 1,000 milligrams of calcium citrate, 15 milligrams of zinc, 5 milligrams of manganese, and 2.5 milligrams of copper showed that osteoporosis could be reversed in older women, without exercise and without estrogen, at the rate of 1.5 percent yearly (*The Journal of Nutrition,* 124[7], July 1994). I hope someone will follow up on this by doing another experiment involving more subjects and varying the amounts of trace minerals used to help assimilate the calcium. I believe nature always has a remedy if we can get to the root cause of the problem.

In the case of osteoporosis, as with other dietary deficiency diseases, there is nothing the body can do in the face of prolonged loss of one or more essential nutrients when dietary or supplemental intake is chronically less than the body needs. As I mentioned before, lower-priority functions are sacrificed to save higher-priority functions when there is not enough of some nutrient to meet the needs of both. At the onset of osteoporosis, a hormone secreted by the parathyroid gland signals the bones to release calcium, influences the kidneys to block calcium excretion, and makes possible the assimilation of any calcium passing through the small intestine. If calcium intake plus calcium on hand are not enough to meet the body's needs, the bones suffer the most.

The best calcium strategy I know is to build up the bone density in the teens and the twenties (both sexes), so the bones will have plenty of calcium to bring into later life. Use a balanced diet including foods rich in calcium, phosphorus, magnesium, and trace elements. Eat green, leafy vegetables or carrots, or use your juicer to get plenty of natural provitamin A, which helps build strong bones. Get on a regular exercise schedule in your teens (make sure you include at least twenty minutes of aerobics and enough weight lifting to put pressure [not too much] on the long bones of your arms and legs), and keep on exercising your whole life long. Get out in the sunshine (direct or indirect) every day for at least twenty minutes to get plenty of natural vitamin D, the "sunshine vitamin," which is necessary for calcium assimilation.

There are too many stories in our time about an aunt or grandparent who has fallen, broken a hip, and died seven or eight months later. Osteoporosis is nearly always the reason why hipbones are broken when older people fall. (Older men get osteoporosis, but not as frequently as women.) Let's put an end to it starting with our generation. Let's put an end to this epidemic of osteoporosis in our time and wipe out calcium deficiency in our families!

CAUSES OF CALCIUM DEFICIENCY

The primary causes of calcium deficiency are inadequate intake of calcium-rich foods, lack of vitamin D, and magnesium deficiency. Secondary causes include insufficient intake of vitamin A, manganese, copper, zinc, and lack of exercise. Low levels of estrogen or testosterone may reduce calcium assimilation. High magnesium intake or high blood levels of phosphorus blocks assimilation of calcium. Caffeine drinks with sugar accelerate urinary loss of calcium in all ages. Sprue (celiac disease) causes extreme calcium deficiency if not treated promptly.

SYMPTOMS OF CALCIUM DEFICIENCY

In children, bone deformities, soft flexible bones (rickets), and stunted growth result from calcium deficiency (or vitamin D or phosphorus deficiency). Early signs of adult deficiency are tingling or stiffness in hands or feet, cramps, and spasms. Later signs are soft bones and bone pain (osteomalacia), brittle and porous bones (osteoporosis), and failure of blood to clot. Aspartic acid deficiency always accompanies calcium and magnesium deficiency. (Aspartic acid is involved in DNA and RNA metabolism and in liver function.) Low blood calcium in advanced calcium deficiency can result in convulsions and great pain.

Adult RDI: 1,000 milligrams

Best Calcium Foods: Milk, yogurt, cheese (lactose-free products are now available for lactose-sensitive persons), goat's milk, soy milk, canned salmon, canned sardines, steamed leafy greens (kale, spinach, dandelion, mustard), broccoli, tofu, figs, legumes, and raw nuts and seeds (also their butters and milks).

Best Supplements: calcium citrate, calcium carbonate, and calcium lactate.

PHOSPHORUS: THE LIGHT BEARER

I call phosphorus "the light bearer" because that's what it means in Greek. "Phos" means light and "phoros" means carrying. I have been on passenger liners and on small boats in the ocean at the time of year when millions of phosphorus-containing microorganisms were on the surface of the sea. As the bow of the ship or boat plowed through the sea in the night, a lovely luminescent light curled away from each side of the boat in the white froth and reappeared in the wake behind us, glowing in the dark. Phosphorus is certainly true to its name—"light bearer."

Phosphorus does not occur in a free state but in the form of phosphates and alkaline salts. It is in the bones in the form of calcium and magnesium phosphate (where it does not glow in the dark) and is an important electrolyte as well. Blood concentrations of phosphorus and calcium reveal a teeter-totter effect—if one is up the other is down. The body contains about 800 grams of phosphorus at any particular time. It fluctuates in its interaction with calcium and requires the help of vitamin D to be assimilated from the small intestine. Seventy to 80 percent of this stored phosphorus is in the bones and teeth, 10 percent is in muscle tissue, and the rest is in the blood, the cells, the fluid surrounding the cells, and in the nerves and brain. Phosphorus, like calcium, is needed by every cell in the body.

The minerals in bones are completely replaced about every seven years, being deposited and withdrawn many times, just like money in a bank account. The body is designed so that minerals and trace elements that play key roles in body activities have a backup reserve in case of emergency. Extra iron, if needed, is stored in the liver, spleen, and bone marrow as ferritin, an iron-phosphorus protein. Sodium reserves are in the bones, stomach walls, and joints. The liver stores a year's supply of vitamin B_{12} in case of temporary deficiency in the diet. Fat tissue in the body is also an energy fuel reserve. The body is prepared to deal with short-term deficiencies of most essential nutrients.

Phosphorus plays the starring role in many body functions. As a key ingredient of the energy production process in every cell of the body, adenosine triphosphate helps transform glucose into energy and carbon dioxide. Most enzyme reactions involving B-complex vitamins as cofactors can only take place in the presence of phosphorus. As an essential part of the nucleic acids DNA and RNA, it influences cell reproduction and protein formation. Phosphorus helps transport and break down fats. A chemical called phosphocreatin energizes muscle contractions. Lecithin, which contains phosphorus, helps keep cholesterol in solution so it can't deposit on arterial walls and cause cardiovascular disease. Male seminal fluid is mostly lecithin. Lecithin helps substances pass through cell membranes and participates in breaking down fats.

About 70 percent of the phosphorus in foods is assimilated into our bodies, unlike calcium, of which only 20 to 30 percent is absorbed from food in the small intestine. Excess phosphorus and magnesium in the blood hinder absorption of calcium from food. (Calcium, in turn, hinders absorption of iron.) If the intake of calcium, phosphorus, or vitamin D is too low, bones don't grow properly.

Phosphorus, in the form of electrically charged phosphate ions, has a significant influence on water balance and osmotic pressure in the body. Phosphate in the blood helps maintain the acid-alkaline balance. An acid phosphate (monosodium phosphate) works with an alkaline phosphate (disodium phosphate) to stabilize this balance.

Healing of broken bones, rickets, and osteomalacia is speeded up when there is sufficient phosphorus working with calcium and vitamin D.

CAUSES OF PHOSPHORUS DEFICIENCY

Antacids with aluminum in them block phosphorus intake, as will an excess of iron. Lack of vitamin D or a high blood level of calcium will block phosphorus assimilation. The hormone calcitonin causes rapid loss of phosphorus, and sugar upsets the calcium-phosphorus balance. Alcohol also interferes with phosphates. Phosphorus is in so many foods that it is seldom insufficient in the diet.

SYMPTOMS OF PHOSPHORUS DEFICIENCY

Early symptoms would be loss of appetite, weakness, weight loss, and bone pains. Later, bone softening and weakening, bone malformation, and poor development of teeth would mean rickets was setting in.

Adult RDI: 1,000 milligrams

Best Phosphorus Foods: Meats, fish, poultry, milk products, eggs, caviar, codfish roe, legumes, cereal grains, bran from cereal grains, squash, soybean products, and sprouts. (One cup of dried garbanzo beans has 732 milligrams of phosphorus, and 1 cup of crude rice bran has 1,392 milligrams of phosphorus.)

MAGNESIUM: THE RELAXER

I call magnesium "the relaxer" because it stimulates the relaxation phase in muscle tissue, including the heart muscle, just as calcium stimulates the contraction phase. Many people are familiar with milk of magnesia, an antacid suspension of magnesium hydroxide sometimes used as a cathartic, and with Epsom salts, a bath salt of magnesium sulfate that produces wonderful relaxation. Dr. C. Norman Shealy considers magnesium "a crucial neurochemical," which I will discuss in the next few pages.

In an average person, there are close to 25 grams of magnesium, 70 percent of it utilized in the bones, and the other 30 percent used in the soft tissues and intracellular fluid inside our cells. There is very little in the fluid surrounding the cells.

Magnesium is one of the key chemical elements in the operation and energy production of human cells.

Magnesium stands out among the chemical elements for three big reasons. First, it takes part in over three hundred enzyme reactions, which means it has enormous influence in the body. Second, it works with adenosine triphosphate in cells to produce energy, the critical factor in overall body metabolism. Third, magnesium deficiency either precedes or follows nearly every disease I know about. In my work as a clinical nutritionist, I have found that all illnesses and diseases involve chemical deficiencies. It is only in the past twenty years or so that researchers have collected evidence that magnesium deficiency is involved in just about every significant disease, physically and mentally, that affects both genders and all ages. Many researchers are looking into what can be done to respond nutritionally when people develop magnesium deficiency and food sources are not being well assimilated.

Considering its normal activities in the human body, magnesium activates enzymes, assists in regulating the heart, helps synthesize nucleic acids and proteins, stimulates formation of urea (a breakdown product from proteins), and converts body fuels to energy in cells. It helps transport substances across cell membranes, joins with other electrolytes in regulating the acid-alkaline balance, and works with vitamins C, E, and B-complex in the body. Magnesium serves as an active check on calcium in the body by aiding in its assimilation from the bowel when it is lacking or by stimulating release of the parathyroid hormone, which causes calcium to be deposited on bones when there is plenty in the blood.

In the series of steps converting sugar (glucose) to energy in cells, nine of the most important enzymes in the process require magnesium, sometimes in partnership with phosphate molecules. Magnesium is active in enzymes needed to make adenosine triphosphate, which is at the heart of every cell's energy production system.

Toxic levels of magnesium are seldom encountered because excess magnesium is usually excreted by the kidneys. But if there is a kidney problem, an excess of magnesium develops when food magnesium, supplements and antacids, or cathartics containing magnesium are all being taken on the same day. The antidote for excessive levels of magnesium is calcium. When calcium compounds are taken without magnesium, they cause existing magnesium levels to go down. Special care has to be taken when kidney disease is involved at the time the overdose is discovered.

DISEASES THAT CAUSE MAGNESIUM DEFICIENCY

When magnesium deficiency takes place, no matter what the reason, many of the body's processes that depend on magnesium are curtailed, impaired, altered, or perverted. Deficiencies then allow diseases or dysfunction to settle in, especially in the genetically weak tissues of the body. According to the *Medical Sciences Bulletin* of February 1995, sufficient magnesium in the body "has several antiasthmatic actions: as a calcium antagonist it relaxes airway smooth muscle and dilates bronchioles. It inhibits cholinergic transmission, increases nitric oxide release, and reduces airway inflammation by stabilizing mast cells (which normally release tissue-irritating histamines) and T-lymphocytes." Whether this triggers an asthma attack in someone who has previously had episodes of bronchial asthma or inflicts the very first asthma symptoms on someone who has never had asthma doesn't matter. It is still a magnesium deficiency symptom.

In this section, I list diseases and conditions known to cause or accompany magnesium deficiency. My purpose is to show you that one disease can cause others by creating deficiencies that open up a person to other diseases. The following conditions are known to contribute to magnesium deficiency: Bartter's syndrome; bile insufficiency; celiac disease; bowel infections; vomiting; diarrhea; alcoholism; diabetes; high levels of diuretics, vitamin D, or zinc; hyperthyroidism; metabolic disorders; hormone disorders; fat metabolism problems; colostomy; and kidney dysfunctions.

I want to point out here that the classical deficiency symptoms for magnesium include neuromuscular signs, such as tremors, weakness, muscle spasms, and irregular heartbeat; gastrointestinal signs, such as nausea and vomiting; and personality changes that display confusion, apprehensiveness, and depression. In the "old days," people with magnesium deficiency were often (mistakenly) taken to mental institutions because they acted so radically different that they literally seemed to be mentally ill. There are further consequences once magnesium deficiency has begun.

Most magnesium deficiencies in the early stages develop less dramatically than I've just described. But once they are in place, a whole host of other problems may begin to emerge. Whether the following problems were somehow latent in the body to some extent, and magnesium deficiency was the fuse that set them off, I don't know. I believe that genetic weaknesses in the body almost always play some role in symptom emergence. Conditions that may surface in connection with magnesium deficiency include aggressive behavior, asthma, chronic fatigue, kidney stones, blood clots in the heart or brain, heart attacks, cardiac arrhythmias, hypertension, preeclampsia, migraine headaches, dementia, and osteoporosis. In such cases, magnesium administration by injection, intravenous administration, or oral supplements (whichever is most appropriate to the specific condition) often brings relief—not instant healing of the condition but at least temporary relief. What interests me at this point is how far we can go to remedy this situation by paying more attention to magnesium foods in the diet.

Body chemistry, to me, is one of the most exciting, interesting, and promising approaches to health care in our time. The main point of bringing the subject of magnesium deficiency out in this way is so people like you will be interested and talk to others about it. This is a nutrition issue that has unavoidably become a medical issue because of the far-reaching health consequences it entails.

Adult RDI: 400 milligrams

Best Magnesium Foods: Raw rice bran, raw wheat germ, yellow cornmeal, corn, soybeans, soy milk, tofu, raw seeds and nuts, seed and nut butters, seed and nut milks, all green, leafy vegetables, all yellow vegetables and fruits, whole cereal grains, milk products, and seafoods. Meat and poultry are not particularly good sources of magnesium.

THE VITAMIN D FAMILY

Vitamin D is a family of steroid hormones, not really a vitamin, but because it was originally labeled a vitamin in 1921 by its discoverer, Sir Edward Mellanby, it retains the title. There are now five variants of vitamin D, all of them fat soluble, all of them equally effective in assisting calcium and phosphorus through the bowel wall and into the bloodstream. We can get vitamin D from fish liver oil, from spending twenty minutes in the sunshine, or from drinking vitamin D–fortified milk. The vitamin D hormones that enter the body stay inactive and useless until they have traveled first to the liver for a change, then to the kidneys for another and final change into their active form.

Vitamin D is essential for bone formation. Without vitamin D, calcium and phosphorus could not be assimilated into the body or deposited onto the bones. In centuries past, deficiency of vitamin D caused a bone-softening disease called rickets in children and osteomalacia in adults. In this country, milk was fortified with vitamin D in the first half of the last century to put an end to the deficiency diseases, and it almost succeeded. Doctors still run into a case every now and then.

Sunscreen prevents vitamin D from being synthesized by blocking ultraviolet light from altering the sterol carried by tiny blood capillaries just under the skin. If you don't think you're getting enough vitamin D, you can buy it at your local health food store.

CAUSES AND SYMPTOMS OF VITAMIN D DEFICIENCY

Lack of sufficient vitamin D causes faulty bone mineralization as previously described. The symptoms of vitamin D deficiency are the same as for calcium deficiency—tingling, muscle spasms, and numbness. Using mineral oil, barbiturates, or cholestyramine, or abusing alcohol can cause vitamin D deficiency. So can liver or kidney disorders, or certain bowel problems. If you are deficient, see your doctor to find out why before buying a supplement.

SYMPTOMS OF TOXIC EXCESS

Acute overdose is indicated by lack of appetite, feeling nauseous, frequent urination, weakness, fatigue, diarrhea, and vomiting. Advanced symptoms for long-term excess could include disorientation, hypertension, kidney failure, and coma. Infants at the stage where they can drink milk from a glass are sometimes oversensitive to the vitamin D in fortified milk and get hyperactive. Call your pediatrician if this happens to your child. There is no good reason for adults to take more than 400 IU of vitamin D daily unless under a physician's care.

Adult RDI: 400 IU

MANGANESE, COPPER, AND ZINC

Manganese, copper, and zinc were the three trace elements used together with 1,000 milligrams of calcium citrate malate to reverse calcium loss due to osteoporosis in a group of postmenopausal women (mean age sixty-six years) and to replace calcium at the rate of 1.48 percent over a two-year period. The amounts of trace elements used daily were manganese, 5 milligrams; copper, 2.5 milligrams; and zinc, 15 milligrams. The study was supervised by Linda Strause, Paul Saltman, and others from the departments of Biology and Community and Family Medicine at the University of California at San Diego. The results were published in *The Journal of Nutrition* (124 [7], July 1994: 1060–1064). This journal article cited previously published articles describing the effects of manganese, copper, and zinc on the bones of chickens and pigs dating back to the 1950s and 1960s. The earlier studies indicated that trace mineral supplementation along with large dosages of calcium might reverse the bone loss of human beings with osteoporosis. They tried it, and it worked. The researchers who supervised the experiment believe that these three trace elements activate enzymes that influence the restoration of calcium onto the bones of those with osteoporosis.

SLOW-RELEASE FLUORIDE REVERSES OSTEOPOROSIS

Another experimental attempt to reverse osteoporosis was supervised by Khashayar Sakhaee, M.D., professor of internal medicine at the Center for Mineral Metabolism at the University of Texas, Southwestern Medical Center at Dallas, Texas, who tested a slow-release fluoride in the amount of 25 milligrams together with 800 milligrams of calcium citrate twice daily on 980 volunteer elderly subjects with osteoporosis. The results of the three-year experiment were very encouraging. Dr. Sakhaee reported an average gain in spinal bone density of 5 percent per year every year for the three years of the experiment.

In previous experiments, a faster-acting fluoride was used with a calcium supplement in an attempt to treat osteoporosis in volunteer subjects. Over the course of the experiments, the rate of calcium loss from the bones was slowed down but not stopped or reversed. Dr. Sakhaee claimed that administering a slower-acting fluoride with the calcium and having the volunteers take smaller doses twice a day were the reasons for the excellent results with the volunteers. Other physicians have questioned the validity of fluoride supplement restoration of bone on the grounds that it produces a poorer quality of bone. I'm sure this issue will be settled with further studies.

CONCLUDING THOUGHTS

I'm so glad that the experiments just described showed that there was a nutritional way to reverse osteoporosis. Nature cures, but sometimes she needs a helping hand. The problem with drugs that help so much is that they always seem to have undesirable side effects and sometimes long-term effects that don't show up for years. Drugs are trade-off cures; for the privilege of having a few more years of symptom-free living now, we pay later by having to endure side effects that might be much worse than the original disease. I would rather take my chances with what nature has to offer.

In the next chapter, we'll meet the team of hardworking players that build up the blood and enable it to do such a wonderful job of transporting oxygen and nutrients to the cells and getting rid of metabolic wastes. You'll find out that we must have a clean bloodstream in order to have a clean body.

questioned the validity of thyroid supplement restriction of dose on the grounds that it produces [illegible] quality [illegible]. [illegible] this issue will be settled with further studies.

CONCLUDING THOUGHTS

[illegible]

CHAPTER 5

THE BLOOD BUILDERS: IRON, OXYGEN, COPPER, AND COBALT

Not long after I moved into my new health ranch in the mountains above Escondido, California, a family from the Doukhobor community in Canada came to see me. The Doukhobors, who originally came from Russia, are vegetarians, and the parents in my office were concerned about their daughter who had become seriously anemic trying to follow the vegetarian way of life. She didn't want to use any animal products (such as desiccated liver) because that would be a violation of her family's beliefs. They came to me because I had met them during a visit to Canada, and they trusted me. I had lived as a vegetarian, and they believed that if anyone could help her, I could.

I was pleased with their confidence but concerned about whether I could meet the challenge. Iron and vitamin B_{12} are found primarily in meat products, and B_{12} is only in vegetables in trace amounts due to contamination by fertilizer. The girl was pale, tired, and depressed. I told them I would do what I could.

The first thing I did was put her on a diet high in green vegetables and had her drink green vegetable juice—all she could take. I knew that chlorophyll molecules were exactly the same as hemoglobin molecules in red blood cells, except magnesium atoms replaced the iron atoms in the hemoglobin structure. I also knew that iron took part in the formation of chlorophyll and was present in green vegetables.

In only a few weeks, that little girl's cheeks were rosy, her energy was back, and she was outside playing with the cats or feeding grass to our goats. Her red blood cell count was back to normal. Soon her parents came and took her back home with them to Canada.

Anemia is a serious deficiency disease, and people die from it. It is caused by a shortage of red blood cells that normally carry oxygen from the lungs to all the tissues of the body, including the brain. Oxygen is necessary to burn glucose (blood sugar) in the production of energy. The less available oxygen there is, the less energy is available.

The brain and the heart are the two biggest users of oxygen in the body, and iron is the single most important chemical element for transporting oxygen from the lungs to the heart, brain, and all the other tissues via the bloodstream. Iron is found in the hemoglobin molecules in red blood cells and in the myoglobin used to store iron in the muscles.

IT'S ALL IN THE BLOOD

The average 70-kilogram person circulates about 5 liters of blood, which is made up of 35 to 45 percent blood cells and

55 to 65 percent plasma. The plasma is 92 percent water and 8 percent solids. The solids include proteins, nutrients, vitamins, electrolytes, hormones, and metabolic wastes.

The blood cells all originate in bone marrow. Each milliliter of blood contains 4.5 to 6 million red blood cells that carry oxygen, 150,000 to 300,000 platelets that help form blood clots, and 5,000 to 10,000 white blood cells of five different kinds that form part of our immune systems, which protect us against disease. Another part of our immune systems is made up of the protein globulins in the plasma. Some of the globulins can be changed into antibodies that destroy viruses.

Each red blood cell contains 270 million hemoglobin molecules, and when it reaches the lungs, it picks up over a billion oxygen molecules. Oxygen is then carried from the lungs to the tissues and cells where it is used to create energy. The number of red blood cells in women is 4.5 million per microliter and in men it is 5 million per microliter. A microliter is one-millionth of a liter. (A liter is very close to being a quart.) So we have several trillion blood cells per liter of blood, each cell carrying over a billion oxygen molecules. Blood cells last an average of 120 days, while new blood cells are made at the rate of 2.4 million each second, just enough to replace worn-out blood cells.

IRON

There is 3.5 to 4.5 grams of iron in the body, with 70 percent of it in the hemoglobin (blood) and 10 percent in the myoglobin (muscles). Iron is absorbed by the duodenum and small intestines and carried by the blood into the bone marrow, where it is used to make hemoglobin. Twenty-five percent of the total iron is kept in the liver, spleen, and bone marrow, and 6 to 20 milligrams of iron are needed each day. Iron contents are highest at birth and during youth and lowest in old age. Old age requires at least modest iron supplementation.

The blood requires 5 percent of iron; the liver, 1.44 percent; the lungs, 31 percent; the muscles need 16 percent; the bile uses 13 percent; and menstruation entails a greater loss of iron than any other body function. Brain activity, breathing, cellular respiration, and every activity of the body depend on iron in the blood. Iron (heme) in the hemoglobin of the blood is the oxygen carrier. It promotes blood oxygenation and delivery of oxygen to the cells. Iron combines with oxygen in the presence of moisture in the lungs.

Rosy, delicate cheeks and beauty of complexion depend upon a high-iron diet, but it has to be ferrous iron, Fe^{2+}, to be readily used. The body absorbs this kind of iron, which is found in meat, poultry, and fish, three times more readily than iron from eggs, legumes, and whole grains. Iron in fortified foods may help a little, but not much. We live in an age and a country in which eating habits and lifestyles have led to a widespread epidemic of iron deficiency. Some experts blame this epidemic on the new popularity of "lite" foods and reduction of meat in the diet. About 60 percent of Americans are short of iron.

An excess of iron in the blood causes problems, and we do run into iron overdosing from time to time. Iron excess can develop from continuous blood transfusions, as a consequence of alcoholism, or due to a genetic disorder. Rarely, an individual here and there will overdose on iron supplement tablets. Excess iron damages the liver and spleen.

WHEN A HIGH-IRON DIET IS NEEDED

Iron-deficiency anemia is second only to obesity as a nationwide problem in the United States. Nine out of ten women in the United States are not getting the RDI of 18 milligrams per day of iron. Those at highest risk for iron deficiency are children between the ages of nine and thirteen, premenopausal women, pregnant women, and seniors. I am not saying that we have an epidemic of anemia on our hands. It takes a really severe deficiency of iron to develop a clinical level of anemia. (There are many possible causes of anemia, and iron-deficiency anemia is not the most frequently encountered type in this country.) The problem of iron deficiency manifests primarily as low-level fatigue, lack of energy, and not feeling up to par, so to speak.

Lack of sufficient iron means, first of all, that we are not getting all the oxygen we need to the brain, the heart, and all the other tissues of the body. Thought processes and memory are just not quite what they should be. The heart strains just a little harder to provide normal circulation of the blood. Oxygen is required for energy production in cellular respiration, and when energy production is even slightly impaired, all body and mental functions are slightly impaired. However, not all Americans are only slightly deficient in iron. Many are seriously deficient, and some are anemic.

SYMPTOMS OF IRON DEFICIENCY

Signs of a low-grade deficiency in children are impaired learning capability, attention deficit, constipation, lethargy, and tiredness. The same symptoms apply to adults. A laboratory blood test will show plasma iron of less than 40 micrograms per deciliter

(100 milliliters). Pallor of the face and fingernails, headaches, breathing problems, tiring quickly at exercise, lack of appetite, cold hands and feet and, over the long run, enlarged heart.

Menstrual losses of blood average 28 milligrams monthly. One in four college-age women are iron deficient, possibly due to ignoring the need for increased iron intake prior to and during menstruation. Diseases that cause iron deficiency are colon cancer, hiatal hernia, ulcers, hemorrhoids, bladder tumor, and diverticulosis. Soft drinks high in phosphates cause iron to be excreted in the urine. Lack of copper or manganese in the diet reduces iron assimilation, as does a deficiency of vitamin C. Oxalic acid foods, such as chard and rhubarb, can block iron intake.

FOODS HIGH IN IRON

The following foods contain more iron than most other foods and should be used in preparing iron tonics or blood-building tonics from the other iron foods or juices mentioned. See also Table 5.1)

Artichokes
Asparagus
Blackberry juice
Black radishes
Bone broth
Cabbage, Savoy
Caraway
Collards
Dandelion
Dewberry juice
Dwarf nettle salad
Egg yolks
German prunes
Goat's milk
Goat's whey
Kale
Lamb's lettuce
Leeks
Lettuce
Marjoram
Mustard greens
Okra
Onions, steamed
Rice bran muffins
Rye meal bread
Rye meal muffins
Shallots
Sorrel
Spinach, wilted
Strawberries
Whole rice
Whole rice bread

Table 5.1. **Foods Highest in Iron**

Food	*Milligrams per 100 grams*	*Food*	*Milligrams per 100 grams*
Black strap molasses	16.1	Parsley	6.2
		Egg yolks	5.6
Pumpkin seeds	11.1	Venison	5.0
Wheat germ	9.4	Almonds	4.7
Veal liver	8.8	Raisins	3.5
Chicken liver	7.9	Beet greens	3.3
Black beans	7.9	Chard	3.2
Dulse	6.3	Broccoli	1.1

DOES IT MATTER WHAT KIND OF IRON WE GET IN OUR FOODS?

When evidence of widespread deficiency of a particular vitamin or chemical element becomes verified, government agencies, such as the National Institutes of Health, Center for Disease Control, and the Food and Drug Administration, begin looking for ways to respond to the issue. In the past, widespread problems with goiter led to the addition of potassium iodide in table salt in this country, and the iodine deficiency that was causing the problem with goiter dwindled to a very low level of incidence. (People who don't use salt at all sometimes develop goiter.) When rickets became a big problem, vitamin D was added to most of the milk sold in stores, to assist in the assimilation of calcium. Now that a nationwide problem with iron deficiency has surfaced, iron is being added to flour, breakfast cereals, and breads. But it's not enough.

The problem is that the enriched products are not providing more than 25 percent of the RDI, and if you are not getting the iron you need from your foods, you may have to take a supplement of ferrous fumarate, ferrous gluconate, or ferrous sulfate. Because iron taken by itself interferes with zinc and calcium, you should take it between meals with 100 to 150 milligrams of vitamin C or a glass of orange juice. A glass of orange juice doubles the amount of iron assimilated. A cup of nonherbal tea reduces iron assimilation by 75 percent. You should not be taking iron supplements at all unless you know you are deficient in iron.

Just having anemia doesn't mean you are deficient in iron. Anemia can be caused by a deficiency of vitamin B_{12}, folic acid, copper, manganese, or a substance called intrinsic factor, a specialized protein that transports vitamin B_{12} from the stomach to the bloodstream. If you believe you are anemic, go to your doctor and find out what is causing it. On the other hand, if you know you are not getting enough iron, then you are a good candidate for iron supplements.

One of the best iron supplements is Chromagen, which has ferrous fumarate, vitamin C, and desiccated stomach substance, a stomach protomorphogen that is claimed to improve assimilation of vitamin C. Considering the various iron supplements by themselves, desiccated liver is the best, in my opinion. After desiccated liver, I would recommend ferrous fumarate, then ferrous sulfate, and, lastly, ferrous gluconate. Avoid any supplements whose chemical names begin with "ferric." This is a form of iron with a higher electric charge than the ferrous minerals, and it has to be reduced to the ferrous state before it can be used.

BE CAREFUL WITH IRON

When red blood cells get old (about 120 days), they are scrapped by the spleen or liver and the iron is saved for use by new red blood cells. Small amounts are excreted in the urine and feces, but the largest losses of iron are from bleeding. There is no automatic safety factor in the body to guard against excess. Older persons sometimes store too much iron, which promotes the formation of free radicals, according to Dr. Leo Galland, who points out that some scientists think excess iron increases the risk of cancer and heart disease.

We know that bacteria thrive in iron-rich blood, so we have to be careful of infections. Studies have shown that when iron supplements have been given to Somali nomads or Masai natives in Kenya, their rate of infection increases. Taking adequate amounts of iron doesn't increase the strength of the immune system. Iron, taken at the same time as zinc, will reduce zinc assimilation. Zinc is absolutely necessary for a strong, healthy immune system. That's why I say to take iron between meals, so you don't knock out zinc.

Keep in mind that iron is a key element in the body's energy-producing system. If you lack energy, tire easily, or would like more energy, you should find out if your serum ferritin level is low. That's the best lab test for iron deficiency. Meanwhile, don't take any iron supplements over 20 milligrams. (You can find tablets or capsules with up to 300 milligrams of iron. That's a very dangerous dose.)

OXYGEN

Oxygen, attached to iron atoms in the hemoglobin of the blood, is circulated along with nutrients and chemical elements to every tissue of the body. The oxygen is released into the fluid surrounding the body cells and drawn into each cell together with glucose (blood sugar). The inside of each cell is like a tiny city with many activities going on. The oxygen and glucose are transported inside the cell to microscopic "energy factories" called mitochondria. Glucose is the fuel ignited and burned by the oxygen to produce energy. Energy production is accomplished by a multistep chemical process inside the mitochondria, with the help of special enzymes for each step. (There are several thousand enzymes in each cell.) The job of each enzyme is to produce a single change in the substance being processed, and this can be as simple as adding or removing an electron, or moving an amino acid from one molecule to another, or simply speeding up a process that normally takes more time. The glucose "fuel" is usually obtained by splitting carbohydrates into simpler sugars until the end result is glucose. Proteins and fats can be burned for energy, too, but less efficiently than carbohydrates. Energy not used immediately by each cell is stored in a small molecule with a big name—adenosine triphosphate.

Chronic fatigue syndrome (CFS) is one of the unfortunate by-products of our contemporary culture. Somehow, the energy-producing process I just described doesn't work as well for people with CFS. They may have one or more genetic weaknesses that interfere with energy production. Or there may be a deficiency of one or more of the metal ions that activate enzymes, or a low blood sugar that reduces the amount of fuel available for energy production. There could be a deficiency of copper or phosphorus. When it takes a series of operations to produce something, a flaw in any step of the process could cause trouble.

There is a challenge to the healing arts of our time, and that is to improve the quality of life for every woman, man, and child who comes to us with "something wrong." I'm sure nature has an answer for each problem, even if we haven't discovered it yet. We need to find out how to increase the energy level and well-being for each patient who comes to us, to help her maintain a more robust level of health and live a reasonably long life. In order to do that, we need to make better use of the nutritional knowledge we already have. We need to delve more deeply into how we can cooperate better with nature in producing energy.

When the body is charged with oxygen, the heartbeat is more vigorous, the tissues are more elastic, emotions are animated, and the mind is quick and resourceful. There is no chemical element more vital to health, vigor, and efficient brain function than oxygen. A person can go without food more than a month, and without water for about a week, but without oxygen, a human being dies of asphyxiation in minutes.

OXYGEN ALONE IS NOT ENOUGH

Iron and oxygen are not the only key factors in energy production. Good blood circulation is necessary to get the blood to all parts of the body, which requires regular exercise that involves the heart and lungs. You may enjoy "pumping iron" or using the wonderful machines at the family fitness center. You may like long-distance running, aerobics to music, swimming, bicycling, competitive sports, team sports, hiking, calisthenics, or ice-skating. You have a lot of choices. Exercising for half an hour before breakfast each day raises the metabolic level and keeps it burning more calories all day, as compared with the metabolic level when we don't exercise (or when we exercise at any other time of the day). We need regular aerobic exercise every day to live at peak health.

Sometimes I think one of the most appropriate exercises for Americans would be to push themselves away from the dining table sooner than they tend to do. I feel it is significant that obesity and iron deficiency are the number one and number two health problems of our time in the United States. We can learn an interesting lesson from statistics gathered over the years from 1965 to 1977. People in the United States over that span of years lowered their calorie intake by an average of 10 percent while at the same time experienced a slight gain in average weight. Think about it. Isn't that interesting? We can only assume that they (or we) were physically less active over that twelve-year span, so that some of their calories had to be stored in fat instead of being burned up by exercise. Television may well have had something to do with it.

Exercise is not just another lifestyle option for a person who wants to preserve the quality of life and extend his or her productive years as long as possible. Our bodies appear to have been designed to require a certain minimum level of physical activity—either through work, exercise, or both—that we dare not ignore. Obesity is not a sign of well-being. It is a message from reality that we have trespassed some natural law that applies to our bodies. It is a warning that we are moving into a domain of higher health risk, especially of diabetes, high blood pressure, arthritis, cancer, and heart disease.

Do we understand fully that it is possible to take a disciplined stand on behalf of a healthier, happier, longer, and more productive life by not eating as much as we might like and exercising more than is convenient?

MANGANESE, COPPER, AND COBALT (VITAMIN B_{12})

I call manganese "the love element" because, in a classic manganese deficiency experiment, a mother rat deprived of this trace element ignored her newborn litter and refused to feed her babies. So, I think of manganese as an element necessary for motherly love. I have nicknamed copper the "helper element" because copper so often works together with other elements such as calcium, iron, zinc, nickel, and manganese. Cobalt is a very important element that exists in our bodies only as a part of vitamin B_{12}, and, because it prevents anemia and fatigue, I refer to it as the "antifatigue element."

Manganese

Let's discuss manganese first. The average human body contains 10 to 20 milligrams of manganese, and adults need 5 milligrams daily. Most people find manganese difficult to assimilate, in part because it competes with copper, calcium, and iron for uptake from the small intestine. Its role in blood building may include assisting in uptake of iron from fruit, vegetables, grains, legumes, nuts, seeds, and other nonmeat sources. Also, manganese enzymes are believed to take part in blood cell formation. This element is in many metalloenzymes and also functions as an enzyme activator. Manganese is in superoxide dismutase (MnSOD), an antioxidant enzyme that neutralizes the most dangerous oxidants. It is required for normal bone metabolism, insulin production, and formation of joint material and cartilage. It is essential for the formation of nucleic acids, sex hormones, nerve transmission, mother's milk, and it helps in the utilization of B-complex vitamins and vitamin E. Early signs of deficiency include weight loss, excessively high blood sugar, impaired hearing, tinnitus, dizziness, slowed growth, lowered fertility, anemia, bone density loss, and heart arrhythmia. Chronic deficiency may lead to cardiovascular disease, osteoporosis, and convulsions. Manganese deficiency has been observed in many types of cancer, schizophrenia, diabetes, and rheumatoid arthritis. Foods high in manganese include hazelnuts, chestnuts, buckwheat, legumes, whole cereal grains, green leafy vegetables, beets, and bananas.

Copper

The average adult carries 110 milligrams of copper in the body and needs an intake of 2 to 3 milligrams daily. Doctors and nutritionists run into copper deficiency fairly often, so we need to make sure we are getting enough. We know copper is needed to build red blood cells because lack of it in the diet results in anemia. It is also needed in absorption of iron from the bowel (manganese may also play a part in this), and for recovering stored iron from the liver or muscle tissue. Copper is found in nearly equal amounts in red blood cells. Copper is involved in thyroid metabolism, the immune system, control of cholesterol levels, protecting joint membranes from inflammation, and in cell energy production. Three copper-containing enzymes are important antioxidants—ceruloplasmin, copper thionein, and copper/zinc superoxide dismutase (Cu/Zn SOD). In the previous section, I mentioned the manganese-containing SOD antioxidant, which differs from this one. These antioxidants help protect fatty acids and detoxify free radicals both inside and outside of the cells. Other copper-containing enzymes form and break down hormones, protect iron ions as they are transferred from the liver to the bone marrow to be used in blood cells, and help form neurotransmitters in the brain. Copper is needed in the hair and nails and in skin pigmentation. Copper deficiency symptoms include anemia, weakness, impaired respiration, high blood cholesterol, skin lesions, nerve problems, lowered immune function, loss of bone calcium, and reproductive problems. Too much copper can drive down zinc levels in the body, causing zinc deficiency and impairing immune function. The best sources of copper include liver, lobster, oysters, seeds, and nuts.

Cobalt (Vitamin B_{12})

Cobalt exists in our bodies only in the form of vitamin B_{12} and is toxic to us in its free state. Fortunately, our bodies store about a year's supply of B_{12} (5 to 10 milligrams) in the liver, kidneys, spleen, bone marrow, and brain. Still, the RDI for B_{12} is 6 micrograms daily because of the importance of this vitamin. Livestock animals that have multiple stomachs and chew their cud can consume cobalt-containing grass and hay without harm and convert it to vitamin B_{12}, and we get the cobalt-containing B_{12} when we eat meat and dairy products. Vitamin B_{12} and folic acid are closely related. Both are members of the B-complex family, both are needed to build the blood, and both are necessary to make the nucleic acids DNA and RNA. However, B_{12} is also needed to nourish nerves and to take care of the sheathing that protects many nerve cells. If B_{12}–caused anemia is mistaken for folic acid anemia, and folic acid supplements are used, new red blood cells will be made. The anemia will appear to be cured, because the B_{12} deficiency will be masked. But while this is going on, nerve damage will be taking place. Paralysis begins at the extremities, and if not caught, correctly diagnosed, and treated, it will move to the spinal cord, leading to paralysis and death. Thousands of people died every year from this disease before doctors figured it out and began to handle it correctly. The disease was called pernicious anemia, and the first effective treatment, discovered in 1923, was to give the patient lots of liver to eat. Often the cause was not B_{12} deficiency, but lack of intrinsic factor, a protein that transports B_{12} from the stomach to the small intestine where it is absorbed into the blood. If intrinsic factor is lacking, either very large intakes or intramuscular injections of B_{12} must be given. When B_{12} gets into the blood, it is picked up by transport proteins called globulins and delivered to the bone marrow where blood cells are made. B_{12} assists in the early stages of red blood cell development; then iron, folic acid, vitamin C, and amino acids enter in to finish the job. Copper and manganese assist in the formation of the cells.

The best sources of vitamin B_{12} are liver, other organ meats, meat, poultry, fish, clams, and lobster. The B_{12} in spirulina and chlorella are not assimilated, according to researchers.

LIFE IS IN THE BLOOD

Oxygen-rich blood in the arteries is bright red, while the carbon dioxide–bearing blood in the veins is maroon or brownish red. Blood delivers nutrients and oxygen to the cells and carries away carbon dioxide and metabolic wastes. Like other cells, blood cells use oxygen and glucose to support life, but unlike other cells they are in constant movement, propelled forward by the pulsating systolic beats of the heart through thousands of miles of blood vessels, picking up oxygen in the lungs and dropping off carbon dioxide.

I was surprised when I first found out that 25 percent of the oxygen supply of the blood is used by the brain alone, even though it doesn't move or grow and only weighs about three pounds! Yet, when we take into account that the brain operates all our senses, organs, glands, and muscles, and helps us live and find our place in what we call "the real world," I wonder how 25 percent could be enough. The brain also uses more glucose.

Some of the substances carried by the blood are dropped off at the liver for detoxification or alteration, while other substances, already changed by the liver, are released into the blood for delivery to the cells. As blood passes through the kidneys, its plasma is filtered to remove and discard wastes and excess amounts of sodium, potassium, hydrogen, and other elements present in excess by releasing them in the urine. Urine is 95 percent water and 5 percent dissolved salts and wastes. The kidneys also filter out, preserve, and recycle needed chemical elements, vitamins, amino acids, and glucose back into the cleansed blood plasma to be delivered to the cells. Special hormones signal the kidneys to keep certain substances and get rid of others. The blood and circulatory system in the body are wonderfully equipped to meet the nutrient needs of every organ, gland, and tissue in the body.

In the next chapter, we will introduce and discuss some of the enzymes, hormones, and proteins that use metals and other chemical elements to keep our bodies safe, strong, and productive.

CHAPTER 6

ENZYMES, HORMONES, SPECIAL PROTEINS, AND THEIR TRACE ELEMENTS

Once I was eating breakfast with a business associate and was about to eat a bowl of oatmeal when he said, "I know how to make that oatmeal more easily digested and assimilated in just a few seconds." I said, "You're kidding. How can you do that?" He tore off the top of a small packet and sprinkled a powder over my oatmeal, and in a few seconds the oatmeal turned to liquid. "Digestive enzymes," he said. "Enjoy your breakfast!"

Enzymes, hormones, and what I call special proteins are very important in how they influence our body chemistry, which is why I've devoted a chapter to them. The reason they are so significant is that dietary deficiencies of any chemical elements needed by them usually have a drastic impact on health, tending to do more harm to our bodies in a shorter period of time than deficiency effects on any other body parts. All three are made of the chemical elements I've discussed in previous chapters, but include metallic and sometimes nonmetallic elements. These elements, especially trace elements, often play a key role in most (but not all) enzymes, hormones, and special proteins. By *special proteins,* I mean proteins like hemoglobin, which uses iron to carry oxygen to cells, and ceruloplasmin, a copper-carrying protein that scavenges free radicals and liberates iron from storage sites. We need to know that when we allow key element deficiencies to develop, we are weakening our bodies' natural defenses against infection and disease, and we are interfering with oxygen supplies to the cells. What's the result? Sickness, chronic fatigue, and depression, with higher risk for diseases unless the deficiencies are taken care of.

Metals common in enzymes include calcium, cobalt (B_{12}), copper, iron, magnesium, manganese, molybdenum, nickel, potassium, sodium, tungsten, and zinc. Nonmetals in enzymes, hormones, and other proteins include sulfur (as part of three common amino acids) and phosphate in phosphoproteins, nucleic acids, and proteins called "hormonal second messengers."

At this point, I want to apologize for what I consider the necessary use of scientific names of enzymes and other organic chemical compounds in this chapter. Body chemistry has reached a stage of complexity and depth that demonstrates the extreme importance of developing better nutritional guidelines. I don't expect you to try to pronounce enzyme names or even remember them, but I feel you need to get acquainted with them and what they do. We have to get our feet wet in this river of technical knowledge because it affects our health—yours and mine.

We need to know that 70 percent of all enzymes in living organisms contain metallic elements, and some of the remaining 30 percent require metallic elements that are not actually part of the enzyme to be present to activate them. One-third of all proteins (including enzymes) have trace metals in their structures. Sulfur is in many enzymes, hormones, and special proteins simply because it is a structural part of the amino acids methionine, cystine, and cysteine. Phosphorus is in many phosphoproteins, especially in the nuclei of all cells in the body. For these reasons, I will not be discussing phosphorus or sulfur as special elements. It is sufficient for you to know that both are essential and that people like you and me are seldom deficient in either of them.

I can't say that trace elements are common in hormones, but it is likely that trace elements are in many of the enzymes used for hormone production. Let's start out by looking at what some enzymes are like and what they do.

ENZYMES: THE WORKING-CLASS PROTEINS

Enzymes are protein molecules in plants, animals, and people that make life processes happen by acting as "change agents" in the millions of biochemical reactions that take place in the body every fraction of a second. Enzymes break down molecules and build up new ones. They detoxify poisons and other harmful substances. They act as antioxidants, create energy, protect and repair DNA, make neurotransmitters, and create the fifty thousand different proteins we have in our bodies. They take part in immune system functions, assist in forming red and white blood cells, and help us digest our food. Many scientists have spent or are now spending their lives researching enzymes, and we are far from knowing all that enzymes do.

What parts do the metal elements in them play? They transfer electrons, carry oxygen, act as cell messengers, interact with nucleic acids, and partner with water-soluble vitamins in getting enzymes to do special tasks. They seem to play key roles in activating enzymes, either as part of the actual molecular structure of the enzyme or by bonding to the substance (called a substrate) the enzyme is going to connect with and change.

Enzymes are everywhere. The interesting thing about them is that they make changes without being changed themselves. They can change the rate of body chemical reactions without an outside energy source.

Metallic elements (and sometimes vitamins) function as coenzymes in certain chemical reactions or else they may be required to activate an enzyme that they are not actually a part of.

There are about four thousand enzymes in every cell in the body. They are big molecules, ranging from five hundred times to fifty-six thousand times heavier than a water molecule. If you can imagine every cell of the body as a microscopic factory that creates energy to manufacture a protein or help move a muscle, enzymes are like the workers it takes to run the factory. Each and every enzyme is designed to make a single change in a chemical reaction, whether it's breaking down carbohydrates to create energy, connecting amino acids to produce a special protein, transferring an electron from one molecule to another, or speeding up a chemical reaction that normally would proceed at a slower rate. Let's look at a few examples of enzymes operating in a familiar area of body activity.

HOW ENZYMES HELP DIGEST OUR FOODS

I want to start out by saying that enzymes already existing in the foods we eat, whether cooked or raw, do not help us digest those foods after we eat them. We hear a lot about the nutritional value of enzymes in raw foods, as if those enzymes could join our own digestive juices and help us to digest the very food they came in. Cooking destroys most enzymes, while raw food enzymes operate at a different temperature and pH than our own digestive enzymes. Also, enzymes are proteins, which, like other food proteins, are partly broken down by pepsin in the acidic environment of our stomachs. Once they are partly broken down, their value for assisting in digestion or any other activity is gone, and their only value to us is nutritional—not bad, but not especially important.

Let's discuss digestive enzymes for a moment to provide a glimpse into how enzymes work. As far as I know, none of them have metallic elements in them.

At mealtime when we take our first bite of food, the enzyme ptyalin in our saliva begins to break down cooked starches (or raw carbohydrates). As the food reaches the stomach, it encounters hydrochloric acid and pepsin. Water, simple sugars, salt, and dissolved crystalloid minerals are immediately absorbed. The hydrochloric acid in the stomach, although very strong, has little effect on protein. However, it forms a strong acidic environment in which pepsin breaks down proteins very efficiently.

The stomach churns the food into a mushy mass called chyme, which begins to exit into the upper part of the small intestine. This part of the small intestine is the most acidic, and in that part of the bowel trace elements like copper are most easily assimilated, while zinc is better assimilated a little farther into the bowel where the acid-alkaline balance is more neutral.

Certain enzymes are programmed to be efficient at specific levels of pH (and not at others), so the enzymes that work well at the upper end of the small intestine do not work well at the lower end where the pH is either neutral or tending toward alkalinity. Human enzymes work well at temperatures close to our natural body temperature of 96.8 to 100.4°F (36 to 38°C). (Liver enzymes would operate best at temperatures between 104 and 105.8°F [40 and 41°C]. because the liver functions at a higher temperature than the rest of the body.) Again, I want to emphasize that enzymes are very environment sensitive and appear to be fine-tuned to work well at very specific body locations.

As the chyme is moved along the small intestine, the pancreas releases juice containing digestive enzymes and sodium bicarbonate, which immediately begins to neutralize the stomach acid that has made its way into the small bowel along with the chyme. (Most bowel enzymes need a neutral pH environment to break down foods.) Bile is released into the small intestine along with the pancreatic juice. Bile emulsifies fats to make them easier for lipase enzymes to break down so they can be assimilated through the bowel wall. The protease enzymes finish the job of breaking down proteins that began in the stomach. They split protein molecules into amino acids (that will later be assembled into new proteins with the help of other enzymes). Finally, amylases split carbohydrates into sugars, which are made available to cells to be converted into energy.

Fats broken down into fatty acids, glycerol, and fat-soluble vitamins are then taken through the bowel wall into the lymph system, while amino acids, sugars, and water-soluble vitamins are taken through the bowel wall into blood capillaries. If the enzymes hadn't broken down the foods, they couldn't have been taken into the body and used.

Digestion is a small but critically important part of the work of enzymes in the body. Enzymes help break down any substance in the body that needs to be broken down, and they help build any new substance the body needs. They are essential to our health and well-being in every respect.

ENZYMES AT WORK IN THE BODY

When you read descriptions of what vitamins and minerals do in the body, including my own descriptions of their functions and activities, the benefits described almost always include enzyme activities without actually giving credit to the enzymes. Why? Because all water-soluble vitamins and many minerals and trace elements inside us act as coenzymes or are part of the structure of enzymes, and often you can't separate what the enzyme does from what its trace element and vitamin components do. Deficiencies of trace elements in the diet can prevent enzymes needing those elements from doing their jobs.

There are about four thousand enzymes in each of the fifty trillion cells estimated to make up the average human body. If you multiply four thousand times fifty trillion, you have the approximate number of enzymes in the human body. That means there's an awesome number of chemical reactions in our body cells every microsecond. Some enzyme-driven chemical reactions take place at the rate of several million "events" per second.

Enzymes interact with substances not by bonding to them electrically but by linking up physically. This is required *before* any chemical reaction can take place. What microbiologists call "an active site" on the enzyme molecule must be attached to the substance (substrate) that is to undergo change. Often a metallic element occupies that active site. There is a lock-and-key relationship between an enzyme and the substrate it is altering. The enzyme key fits the lock in only one substance and no others. That's how enzymes can be so specific in their activity, targeting only one substance.

Sometimes a coenzyme—a vitamin, a trace element, or a protein—partners with an enzyme to take part in the lock-and-key pattern of relationship. All water-soluble vitamins—eight B-complex vitamins and vitamin C—function as coenzymes that work together with enzymes. Unless both coenzyme and enzyme are present, no change can be made on the substrate. Metal elements sometimes initiate enzyme reactions. If the metal isn't present, the enzyme doesn't activate.

The importance of much enzyme activity is that it is changing some molecule, peptide, or protein to fit into some sequence of steps or some gradual, ongoing process that contributes to the health of the body.

If the enzyme's job is to change a million substances into some other form, it has to connect and disconnect from each substance before it goes on to the next one to make the changes. Research has demonstrated that some enzymes are able to complete from one thousand to one million lock-and-key events per second, without tapping into any exterior power source and without undergoing any changes.

This should give you a new appreciation of your body's ability to work toward the level of health you desire, if you do your part. I would hope this knowledge motivates you to make sure your body has all the nutrients it needs to fulfill all the tasks that build high-level well-being.

In his book, *Nanomedicine,* author Paul Frietas, Jr., tells us there are over ten thousand different molecules in an adult body. I believe enzymes are involved with the formation of every organic molecule except those that enter the body and remain in their original form. Enzymes themselves are produced by cell processes directed by RNA using blueprints imprinted on the DNA.

Some researchers feel that aging takes place because of the ongoing generation of free radicals and the gradual degeneration of enzyme processes in our cells. When cells divide, the DNA has to make a perfect replica of itself for the new cell. The integrity of that process is supervised by enzymes inside the cell nucleus that inspect the DNA constantly. When they find damage of any kind, the inspecting enzymes call repair enzymes to repair the DNA. The greatest source of DNA damage is from free radicals generated by the chemical changes required to produce energy inside the cell. Most free radicals formed in cells are neutralized or destroyed by antioxidants, including metal-containing enzymes like superoxide dismutase (SOD).

Over a period of time, fluctuations in the antioxidant populations may allow damage to DNA that goes unnoticed. (No detection system is perfect.) Or repair teams are called but don't completely fix the damage. Defective cells replicate themselves, including the defects. Enzymes may grow less numerous and less efficient. Repair crews may not repair as efficiently as they once did. Errors in DNA alter cellular activities, responses, and protein production, and as these errors multiply, the outward expression of such errors appears as signs of aging. This is just one theory of aging among others, but this we know for sure: Even though we all eventually show signs of aging, we can slow down the process through good nutritional habits and a healthy lifestyle.

HORMONES AND SPECIAL PROTEINS

Hormones are substances made in glands, organs, or body parts that are released into the blood to be carried either to all cells or to special tissues. They are designed to chemically stimulate cells or certain tissues to increase or decrease the level of activity or to secrete another hormone. Our energy levels, sex lives, ability to respond to danger, kidney function, calcium metabolism, digestive enzyme release, and our children's growth all depend on hormones. I will give examples of trace elements required by certain hormones when we come to specific metals and their roles later in this chapter.

I'm sure there are many more trace element–containing special proteins than the ones I know about, but this knowledge seems to be on the cutting edge of what's happening in microbiology metalloprotein research these days. There are fifty thousand proteins in the human body, many of them discovered only in the last ten years. We can safely assume that there are more proteins in which trace elements play a critical role than those we know about at this time. I will present the trace elements I know about along with the enzymes, hormones, and special proteins for each metal.

Once we understand the critical nature of the body's need for trace elements, our focus needs to be on the importance of getting the right intake of nutrients to keep our bodies and minds in top condition.

METALLIC ELEMENTS AND ENZYMES

According to current microbiology research, at least 70 percent of all enzymes and one-third of all proteins (including enzymes) contain metal elements. Many times, trace elements are required in a chemical reaction involving enzymes but are not part of the enzyme structure. Metallic (and nonmetallic) elements can be related to enzyme-driven chemical reactions in different ways. Certain enzyme reactions can only take place in the presence of phosphorus, zinc, magnesium, or other elements. In other cases, a B-complex vitamin has to be joined together with a metallic element (cofactor) or with a metal-containing enzyme (coenzyme) to trigger a chemical reaction. The details of such reactions are not required for our purposes here and are generally boring to people without a microbiology background.

According to researchers, there are enzymes in which either magnesium or manganese can set off the desired chemical reaction. I know of one enzyme operating in human cells that might contain any of five different trace elements in its molecular structure.

Superoxide dismutase, more commonly called SOD, is an unusual enzyme that breaks down even the most destructive free radicals in the body. At least four different types of it are now known, involving five different trace elements as parts of the molecular structure. Those metals are iron, manganese, nickel, and, in one form of the enzyme, a combination of copper and zinc. Others may be discovered in the future. The SOD enzyme especially protects the nucleic acids of cells from attack by free radicals. This is very important because DNA damage is one way that cancer can develop in the body.

Now, let's look at the enzymes, hormones, and other special proteins that have trace elements as key structural parts or as activators.

Iodine

I don't know of any enzymes that contain iodine, but the two hormones produced by the thyroid gland require it. Those two hormones are called T_3 and T_4, abbreviations for triiodothyronine and thyroxine, and they are essential for the metabolism of the whole body. No enzymes or other proteins containing iodine are listed in the literature, but an enzyme named thyroid peroxidase puts the two hormones together.

Iron

Iron-containing enzymes called cytochromes participate in the energy production cycle (called the Krebs cycle or citric acid cycle) of body cells. Energy production in cells takes place in special organelles or power plants called mitochondria, and the energy is stored in molecules of adenosine triphosphate (ATP) until needed. There are several mitochondria in each cell. Cytochrome c is very rich in the heart muscle, which uses more oxygen than most other organs. Cytochrome P450 is used in the liver to detoxify foreign substances. Iron-sulfur enzymes include hydrogenases that also take part in cell energy production, while another set of iron enzymes, including catalase and peroxidase, breaks down chemically dangerous hydrogen peroxide, a by-product of the energy process. There are iron enzymes that protect fats from damage due to peroxides. Another iron enzyme, ribonucleotide reductase, is required for making DNA. Iron teams up sometimes with selenium and nickel, and sometimes just with nickel in hydrogenase enzymes, which are able to influence chemical reactions by donating or accepting an electron.

Iron is in two special proteins, hemoglobin, which carries oxygen in the blood, and myoglobin, which stores oxygen in the muscles for future use. Iron itself is stored in proteins called ferritin and hemosiderin in the liver and heart muscle. Transferrin delivers iron to the tissues.

Zinc

Known to take part in over two hundred enzyme-driven chemical reactions, zinc is like a musician who can play any instrument. Zinc, more than any other metallic element, bonds easily to organic molecules. Over 95 percent of zinc in the body works inside the cells—as opposed to hanging around in the fluid outside the cells or circulating in the blood.

Zinc is in each of the following enzymes and plays an important part in the enzyme function. Alcohol dehydrogenase detoxifies alcohol. Carbonic anhydrase deposits calcium on bones and teeth. Carboxypeptidase breaks down proteins. Zinc activates an RNA–synthesizing enzyme and is a cofactor for the enzymes that make and break down collagen.

Enzymes utilizing zinc are said to be active in the production of immune system components. Antibodies are made from proteins called globulins, while white blood cells originate in bone marrow. When the body has to multiply cells quickly for healing or growth, zinc enzymes make it possible. Zinc is especially rich in the brain cells, which would indicate that zinc enzymes are involved in oxygen utilization to energy brain functions. Zinc is important in hormones. Follicle-stimulating hormone, usually just called FSH, contains zinc. Its job is to promote ovarian follicle growth, stimulate estrogen secretion, and develop sperm in male testes. Zinc is in luteinizing hormone (LH), which triggers ovulation and secretion of progesterone in females. It initiates secretion of testosterone in males. Zinc enzymes are critical to the metabolism of insulin, the hormone secreted by the beta cells of the islets of Langerhans in the pancreas.

Metallothionein protein, found extensively in the brain, contains zinc, and so does cysteine-rich intestinal protein (cysteine is an amino acid). RNA proteins contain zinc as part of their structures. RNA, a protein, directs cellular protein synthesis using patterns taken from the DNA in the cell nucleus.

Magnesium

This important nutrient is involved in over three hundred enzyme reactions in the body! Seven magnesium-containing enzymes are required in the conversion of blood sugar to pyruvate at the doorway of the energy production cycle in our cells. We might remind ourselves that energy is an essential ingredient of life itself. The next series of steps in the energy cycle, the conversion of pyruvate to acetyl coenzyme A, requires an enzyme that is magnesium dependent. More magnesium enzymes take part in the Krebs cycle for energy production, but I'm going to back off here just to keep you from going into an attention deficient mode.

Let me sum up. A lot of magnesium enzymes are involved in energy production processes, and, many times, a phosphate-dominated molecule called ATP (adenosine triphosphate) and other cofactors are needed. Magnesium is also involved with enzymes in the metabolism of fatty acids, proteins, and carbohydrates. Magnesium is involved with heart health in the sense that it plays an important part in the interaction of enzymes and neurotransmitters with nerve receptors in the heart muscle.

Magnesium deficiency is relatively common in the United States. In many cases, it exists because of inadequate dietary intake. In other cases, it occurs as a side effect of some relatively common diseases, and a significant number of prescription medications tend to block its assimilation.

Copper

Ferroxidase enzymes regulate iron transport and release. (Copper shortage can cause anemia.) Copper is part of tyrosinase, which changes tyrosine to the hair and eye color pigment melanin and protects skin from ultraviolet light. Cytochrome c oxidase uses cytochrome c as its coenzyme to deliver oxygen to mitochondria (energy factories in cells) and oxidize carbohydrates, fats, and amino acids to create energy. Ceruloplasmin attached to the outside of cell membranes protects fatty acids from free radicals in the fluid outside of cells, changes ferrous iron to ferric iron by stealing an electron, and helps liberate iron from storage. Like ceruloplasmin, extracellular SOD is an effective antioxidant in blood plasma and in the fluid outside of cells.

Copper enzymes catalyze the formation and destruction of the hormones epinephrine (stored in and released from adrenal glands) and norepinephrine (the most important neurotransmitter in the sympathetic nervous system).

Manganese

There is manganese in glutamine synthetase, an important brain enzyme; in arginase, which changes one amino acid into a different amino acid and forms urea; and in pyruvate carboxylase, which catalyzes the first step of the energy-producing cycle in cells. The manganese SOD enzyme, like all the other SODs, is primarily an antioxidant in the mitochondria, our cellular energy factories. Manganese activates large numbers of hydrolases (they split off water molecules), kinases (they steal phosphate from one molecule and give it to another), decarboxylases (they remove carbon dioxide, mostly from amino acid molecules), and transferases (they transfer one or more atoms from one molecule to another).

Molybdenum

Xanthine oxydase has molybdenum in it and is involved in uric acid metabolism and in transferring electrons from one molecule to another. Aldehyde oxidase breaks down toxic purinelike compounds. A molybdenum enzyme named sulfite oxidase stabilizes cysteine (a sulfur-containing amino acid) metabolism and detoxifies sulfite, a food additive banned by the FDA after several deaths and thousands of complaints of breathing problems. It was once used at salad bars to keep vegetables looking fresh.

Selenium

This element is present in some enzymes when it is chemically bonded to the amino acid cysteine and, to a lesser extent, cystine and methionine (all three of these are sulfur-containing amino acids). Sulfur and selenium seem to have an affinity for one another. Selenium is in certain hydrogenases in combination with nickel and iron. Hydrogenases use hydrogen to change the substances they interact with. Selenium is also in glutathione peroxidase, which transfers oxygen from peroxides to tissues, and heme oxidase, which takes away electrons or combines with oxygen.

Chromium

This trace element is a cofactor along with the hormone insulin in regulating blood sugar. Chromium enhances insulin effectiveness by increasing cell sensitivity to insulin. No chromium-dependent enzymes have been discovered.

Nickel

This trace element is in combination with iron and selenium in the hydrogenase mentioned previously in the sections on iron and selenium, which is known to be active in oxidation-reduction reactions (giving or taking electrons), an important activity of a class of enzymes. Nickel also has a role in the regulation of hydrogenases. Nickel binds easily to amino acids and to proteins such as globulins (which can be turned into transporters of metals in the bloodstream). Nickel proteins may be used to stabilize certain molecules.

Vanadium

Vanadium has the unusual ability to imitate the hormone insulin in regulating blood sugar. At this point, little is known about its activity in the human body.

THE CUTTING EDGE OF BODY CHEMISTRY

If I had my way, body chemistry would not be this complicated. However, the bottom line is that microbiology research is revealing more and more details about how the chemicals in our bodies react with each other to keep our bodies alive, healthy, and full of vitality. The more we know about what our bodies need, the more responsible and intentional we can be about creating diet plans that provide the best nutrition we can possibly give ourselves. Metalloproteins are on today's cutting edge of research, and tomorrow our updated knowledge about them may change the way we look at human nutrition and smart food planning for mealtimes.

CHAPTER 7

FOODS AND RELATED HEALTH DANGERS

In a previous chapter I have mentioned the "window rule" regarding vitamins, minerals, trace elements, and other nutrients. That means, for best health results, individual vitamins and minerals should be taken in amounts that fit within a window of upper and lower limits—not so much to cause a toxic reaction or to interfere with other nutrient intakes and not so little to be of no benefit. The window rule also applies to the total diet. We shouldn't eat so much that we are inviting obesity or so little that we become undernourished or at risk for anorexia nervosa. Eating too much or too little leads into an arena of higher risk of disease.

We wouldn't have a problem with dietary excess of vitamins, minerals, and trace elements if our nutritional intake was entirely from foods. However, many people take vitamin-mineral supplements along with their meals because they believe, as I do, that too many foods are grown on depleted soils that lack trace elements. It is possible to overdose on the total daily intake of certain vitamins, minerals, and even amino acids when we add the nutrients in our foods to the nutrients in our supplements. Too much iron can increase free radicals in the body. Too much calcium can lead to tetany, which is a nervous disorder marked by sudden, intense spasms in the arms and legs and possibly swelling. Too much vitamin D can cause headaches, nausea, and kidney stones.

Most foods are not entirely safe, and I am including natural foods in this category. There are toxic chemicals in some foods, such as aflatoxin mold that grows on contaminated peanuts and causes liver cancer. Overuse of ordinary table sugar has been correlated with breast cancer. Overuse of high-fat, high-cholesterol foods increases the risk of heart disease, the number one cause of death in our country. Cancer researchers claim that as much as 30 percent of all cancers are diet related, and almost every food carries some level of carcinogens.

There are food chemicals that block essential nutrients from being assimilated. Phytate in whole grains, wheat bran, beans, nuts, and seeds blocks absorption of zinc and manganese from the small intestine, leading to deficiency problems such as stunted growth and impaired immune system function. Oxalic acid in spinach, chard, almonds, cranberries, rhubarb, and other vegetables prevents calcium from being assimilated. High fiber from foods can bind to minerals and trace metals in the bowel and prevent their assimilation. On the other hand, high-fiber foods lower blood cholesterol and intake of fatty acids and reduce the chances of colorectal cancer.

There are lots of cancer risks around. Unsaturated fats (vegetable oils, liquid at room temperature) increase the risk of cancer. Natural mutagens that can cause cancer are found in alfalfa sprouts, mushrooms, peanut butter, burned toast, and black pepper. Not only charbroiled meat but all cooked meat, fish, and poultry contains carcinogens. Frequent use of red meat in the diet increases colon cancer risk.

Anything chemical, natural, or manufactured that can break or damage the DNA in our cells once it enters our bodies can cause cancer.

EFFECTS OF COOKING ON FOODS

Cooking always causes loss of some nutritional values, but when food is chronically overcooked, not only is folic acid destroyed (eventually leading to anemia) but each meal's nutritional value is diminished. Frying food at high heat develops carcinogens at the interface between the metal frying pan and the food. Frying food in grease at high heat destroys lecithin (which normally keeps fats in solution in the blood) and allows a high concentration of fats and cholesterol to enter your bloodstream. My advice is to get rid of your frying pan.

Some vitamins and minerals are depleted from foods by oxidation, so it is not a smart practice to cut up vegetables or fruits hours before they are to be used and set them aside in bowls. It doesn't help to put them in the refrigerator because it is the air, not the temperature, that is the problem. Boiling vegetables dissolves the water-soluble vitamins, the B-complex vitamins (except B_{12}) and vitamin C.

An article in the *Journal of Home Economics* really caught my attention. It said that boiling foods causes loss of 48 percent of the iron, 31 percent of the calcium, 46 percent of the phosphorus, and 45 percent of the magnesium. I don't believe that it's any accident that iron, calcium, and magnesium are three of the foremost mineral deficiencies among adults in this country.

These three elements are involved with many important enzyme processes that produce toxic by-products if the active metallic atoms are not available, and the same thing can happen when vitamins are boiled away.

I'd like to give an example of how this works. When the B vitamin thiamin is destroyed by boiling or overcooking, it is not available to work as a coenzyme with four different enzymes that require thiamin before they can break down carbohydrates to produce energy. What happens then? The usual series of steps to make the energy is broken, and intermediate proteins from the disrupted process build up to toxic levels in the cells. Other enzymes and macrophages are diverted from other tasks to detoxify the overload of useless proteins.

FREE RADICALS AND CANCER

In the last chapter, I described four of the most powerful antioxidants known to researchers—iron, manganese, nickel, and a copper-zinc combination—all of them variants of the enzyme SOD (superoxide dismutase) and each variant containing a different key metal element at the active site of the enzyme. One or two of these SODs carry out hunt-and-destroy missions against free radicals in the fluid outside of cells, but all of them patrol the mitochondria (energy factories) and nuclei inside the cells, scavenging the dangerous superoxides (O_2) released in the energy production cycle and protecting the DNA in the nucleus. This is a cancer prevention operation, because if the free radical superoxide succeeded in breaking the DNA molecule, mutation would take place during cell multiplication. In a similar way, other antioxidants, such as vitamin C, vitamin E, beta-carotene, and selenium, destroy free radicals and protect against mutation, which is sometimes the start of cancer. Our food sources of polyunsaturated fatty acids easily bond to oxygen and form free radicals, but antioxidants move in and neutralize them. Carotenoid antioxidants come from carrots, apricots, peaches, mangoes, papayas, cantaloupes, sweet potatoes, spinach, and parsley.

Other cancer-causing agents are smoked meats and fish, pickled foods, and salt-cured foods. Stomach and esophageal cancers are common in countries like Iceland and Japan where people favor such foods. Heavy alcohol consumption greatly magnifies the risk of cancer in the upper gastrointestinal tract, and cigarette smoking combined with excessive drinking increases the risk dramatically.

Chronic overdrinking or binge-drinking alcoholism causes deficiencies of folate, thiamine, pyridoxine, vitamin A, and zinc, the combination of which causes multiple risks for disease, but mostly for cancer.

GENETICALLY INHERITED PROBLEMS

Everyone is born with some genetic defects, and often these are relatively trivial and do not interfere with normal living. Doctors casually refer to patients as having a "weak stomach," "poor kidneys," or even a "delicate constitution," when they are actually referring to organs, tissues, or systems we have inherited that don't work as well as normal ones.

Heredity plays a part in how our bodies respond to the foods we eat and the diseases we are most vulnerable to. With regard to foods, we may have problems with abnormal digestive enzymes or an inability to assimilate food particles after they have been digested. All such problems are linked to a chain of consequences that produces not just one but many abnormal chemical reactions in the body.

In my own experience with genetically impaired patients, inherently weak organs, glands, and tissues act as if they were metabolically underactive. They have more difficulty accepting nutrients and are not as efficient in getting rid of metabolic wastes as normal tissue. To compensate for this, I designed diets that provided extra nutrients to the underactive tissues, and I taught such patients how to make their eliminative processes more efficient (i.e., exercises to move the lymph, more fiber in the diet, increased intake of water, bowel cleansing, and supplementary herbs).

A person with a genetically impaired digestive system has to take more responsibility for eating correctly to avoid complications in nutrient intake and impaired health.

Celiac disease, now called sprue or gluten enteropathy, is probably due to a genetic flaw that entails immune system dysfunction. The people who get this disease have bowel reactions to the chemical gliaden in the gluten of wheat, barley, rye, and oats, with inflammation and mild-to-severe damage to the small intestinal wall. Symptoms include irritability, abdominal distress and distention, vomiting, diarrhea, terrible gas problems, and weight loss.

White blood cells attack the bowel wall and, in extreme cases, destroy the villi—the microscopic fingerlike projections in the bowel wall that take in food particles. Enzymes in the bowel wall diminish, the internal chemistry of the bowel is compromised, and multiple nutrient deficiencies develop as the digested but unassimilated food is propelled through the intestine.

The only known solution is to remove all gluten-containing foods from the diet. Other diseases involving some degree of reaction to gluten in grains, flours, and other grain products include a type of dermatitis that only men seem to get, Type 1 diabetes mellitus, ulcerative colitis, thyroid disease, and deficiency in immunoglobulin A. The bowel returns to normal and the symptoms disappear when gluten is avoided.

FOOD ALLERGIES AND SENSITIVITIES

Somewhat related to sprue are the food allergies doctors encounter in their work. I have counseled patients who were allergic to just about everything except carrot juice or, in one case, raw goat's milk. True food allergies cause the immune system to react and attack certain foods almost as if they were bacteria attempting to invade the body. I hope you noticed I said "almost."

The antibodies that go after germs destroy them. The antibodies that go after food allergens, immunoglobulin E, stop short of direct contact with the food. They just spray the offending food particles with chemicals, which create inflammation in the tissues. Lips, tongue, and other parts of the face may swell up and tingle. Breathing may become difficult. This is an anaphylactic reaction that can be life threatening in extreme cases. Most of the time, symptoms are not severe.

Runny noses, rashes on the forearms and elsewhere, hives, sinus reactions, migraines, and possibly diarrhea may result. These are the mild but annoying allergy symptoms that most people with allergy problems experience. Only 2 percent of the population have true food allergies.

However, there are food intolerances that are not true allergies. To pick an obvious example, milk intolerance due to absence of lactase is not an allergy. The immune system isn't involved. Sometimes babies are intolerant of milk without being either lactose intolerant or allergic. Sometimes they do well with soy formula. When I started my practice, I found that infants intolerant of cow's milk often accepted raw goat's milk without any problems, but these days raw milk is not advisable for babies because of the possibility of contamination.

Extensive testing of children who reacted to certain foods has shown that if the food is powdered and put into a capsule that the child can't visually identify, there is often no antagonistic reaction. In other words, some food intolerances seem to have a psychosomatic side to them. This is true of many adults as well as children. The best thing to do is cut those foods out of the diet, no matter what is causing the intolerant person's symptoms.

I don't believe we've heard the last word on food intolerance. I believe that enzyme deficiencies affecting food tolerances are far more common than we realize, and I also believe we will see new breakthroughs in this area.

DRUG INTERACTIONS WITH FOODS

This is an area that we can only touch on here because it involves several thousand drugs that have not yet been checked out for their effects on foods, digestion, and assimilation. It should include chemical additives in processed foods and substances like sulfites (now banned from use on salad bars) and monosodium glutamate (MSG), a flavor enhancer used often in Chinese restaurants to which some people react strongly.

Let's start with MSG first. Studies have shown that not everyone who claims to be allergic to MSG is, in fact, allergic. What we have discovered is that MSG affects many normal people who eat their food faster than the body can break down the MSG. When this happens, the body converts the unprocessed MSG to GABA, a neurotransmitter. Both MSG and GABA are able to cause neurological distress in normal people. Of course, some persons are more sensitive to these chemical reactions than others, which accounts for the fact that not everyone experiences unpleasant symptoms from MSG.

For some time, we've known that antibiotics like tetracycline and erythromycin destroy bowel bacteria and block vitamins K, B_{12}, and other B vitamins produced by beneficial bowel bacteria. Antidepressants, such as imipramine, and antipsychotics, such as chlorpromazine, interfere with the B vitamin riboflavin. Seniors who take mineral oil as an aid to regularity are wiping out fat-soluble vitamins such as A, D, E, and K.

Diuretics, by increasing kidney output, deplete calcium, magnesium, potassium, and zinc. This, in turn, depletes energy and lowers immune function. Over-the-counter antacids, as well as the peptic ulcer medication cimetidine, reduce iron and calcium assimilation, vitamin B_{12}, and intrinsic factor (without which B_{12} can't be absorbed). Drugs prescribed for heart disease reduce libido, and some of them render men impotent.

Drugs mixed with other drugs can do unpredictable things in the body, blocking enzyme functions, stressing the heart, damaging the kidneys, and preventing assimilation of nutrients. Anticoagulants, such as Coumadin, can cause chronic diarrhea, which may result in dangerous losses of electrolytes. The eighty-year-old father of one of my employees had terrible diarrhea for three months while taking Coumadin. At the insistence of his wife, the man's doctor checked all the medications he was taking and found that he was over a year past the time he should have stopped using the Coumadin. This drug can be dangerous.

If you want to pursue this kind of search, there are books in your local library or bookstore that can tell you whether the medications you are taking will interfere with your dietary needs.

THE BOWEL AND BODY AS A CATCHALL

Every drug you use will leave residues of some kind in your body even after you quit taking it. While you are using prescription and/or over-the-counter remedies, these drugs will mix with the chemical food additives you take in when you eat processed foods, together with bile salts, pancreatic enzymes, putrefying proteins, alcohol (if you drink), the metabolic by-products of bowel bacteria, and whatever else is there. The bowel, lymph, bloodstream, and liver can become a kind of catchall system, not necessarily able to detoxify or excrete all the chemically harmful substances that interact with one another as well as with the body chemistry. Often, toxic substances are stored in body fat, genetically weak organs and tissues, and the bones.

Tissue cleansing to get rid of all this repugnant debris is one of the best approaches to developing a healthier body. I put my patients on a diet that is half cleansing (getting rid of stored toxins) and half building (repairing damaged tissue). What you have done with your body in the past will determine how hard you will have to work to cleanse and rebuild your body.

In the next chapter, we will wrap up the obvious evidence that knowing your body chemistry is one of the essential keys to being able to do what is necessary for good health and vital energy.

CHAPTER 8

LET YOUR FOOD BE YOUR MEDICINE

Hippocrates said, "Let your food be your medicine, and let your medicine be your food." In my opinion, the kitchen is more important to our health than the doctor's office or the local hospital. If food selection, preparation, and cooking are done wisely, intentionally, and properly, very little time will need to be spent visiting health-care facilities. In this book, I have presented practical information that should give you a basic understanding of how the body functions and what it needs nutritionally to give you the best life you can possibly have. Of course, the mind and spirit are as necessary as the physical dimension to fully describe what I have called the "best life," but this is not the place to bring out the tremendous influence on health that comes from developing the mind and nourishing the spirit. Almost needless to say, the patient needs to be cared for emotionally and spiritually as well as physically.

I have read that what we will eventually die from is imprinted on our DNA. That makes sense to me, although what we bring to life may make an important contribution to how long and how well we live. Certainly our genetic inheritance defines dominant pathways that eventually expose the weakest links in our bodies' systems no matter how we live.

However, a person with a weak constitution and many genetic flaws who follows a nutritionally balanced diet, lives a healthy lifestyle, and maintains a positive attitude may live much longer than someone with a strong constitution and few genetic flaws who ignores nutritional principles, lives for pleasure and self-gratification, and displays an attitude of mixed negative, positive, and whimsical outlooks. How well we take care of the bodies we were born into plays a large part in our health and longevity.

The old saw "what you don't know can't hurt you" just isn't true when it comes to body chemistry, nutrition, and a healthy lifestyle. Ignorance leads to poor nutritional choices, chemical deficiencies, and the kinds of diseases that prey on an undernourished body and shorten the life span. Heart disease, cancer, and diabetes all take a terrible toll on life in this country, and all three of them are strongly influenced by food patterns and lifestyle habits. Hepatitis C, a chronic, incurable version of hepatitis B, is rising up in our nation like a plague that has arisen as a result of cultural factors rather than genetic influence, but diet and nutrition still significantly affect the quality of life for those who have it, and that is true of most physical dysfunctions. The right kind of knowledge can always lead us to a better life, no matter what physical or mental handicaps we have.

In Third World countries, among the urban poor and seniors in this country, and with some who follow the classic macrobiotic diet (and other diets), kwashiorkor (protein deficiency) may become a serious problem. When protein intake is insufficient, amino acids needed for replacement or repair of the fifty thousand proteins of the human body are lacking, including thousands of enzymes and many hormones. Growth of children is impaired and healing of wounds and damaged tissue in adults is slowed. There are changes in the skin, hair, and nails. Loss of appetite, nervousness, irritability, fatigue, water retention, and diarrhea develop. Most of these symptoms are caused by blocked enzyme reactions, but fatigue may be a result of thyroid and insulin hormone deficiencies.

Many reasons could explain why most people eat and live the way they do, but lack of knowledge is generally the foremost reason in this country. It simply doesn't make sense to put our lives into the hands of health educators and practitioners who have never studied nutrition thoroughly enough to offer competent advice to the family. If we have been following stereotyped food patterns passed down through our family for generations, or following cultural traditions about what and how we should eat, it is never too late to study body chemistry and follow a higher pathway to better health.

My heart's desire is to share what I've learned about health in ninety-two years of living and seventy years of professional practice in the healing arts. This includes many years working in sanitariums with "live-in" patients. When you live in the midst of the patients you are trying to guide back to health, your patients become living textbooks. I have always taken it very personally when I encountered patients who didn't seem to be progressing as much as they should, and I would lie awake at night thinking, "What more can I do for this patient?"

Perhaps the hardest thing I faced was the resistance some patients showed toward learning a healthier way to live. Some people *want* to have doctors take care of them. They don't want to take responsibility for their own lives. An artist friend of mine carved a wooden sign for the front of my office building that read, "You're looking for a good doctor; I'm looking for a good patient." To me, good patients are people who are eager to learn how to live so that they will never again experience the same disease or ailment they had when they first came to a doctor for help. Doctors need to do more educating and less medicating, but that only works when patients are motivated to learn how to take care of themselves. And, when you get down to patient realities, there is a very strong emotional and mental dimension attached to foods and eating.

Taking favorite foods away from someone who has been eating and loving them for twenty, thirty, or forty years and teaching them instead to enjoy slow-cooked whole cereal grains, raw vegetables, fresh fruits, lots of garden salads, herbal teas, fresh juices, and raw seed and nut butters, with small portions of meat, fish, or poultry three times a week or less, is not an easy job. There are arguments, disagreements, people defending their right to eat what they love. I tell them I don't blame them and encourage them to make the change because it's the right thing for them to do.

I know. I've been there myself. I was raised on delicious Danish pastry baked by a Danish mother who was a wonderful cook, and I worked my way up to 20 cups of coffee a day. This started in my childhood and continued for many years. But I didn't know any better.

My health broke down in my early twenties, and I developed a lung infection that had proved fatal for others. In those days, no antibiotics or even sulfa drugs were available to treat infections. I didn't have much to work with in terms of bodily health until I ran into a Seventh-Day Adventist doctor who introduced me to the healing power of nutritious foods. I was able to overcome my health problem by simply changing to a better diet.

Diet doesn't stand all by itself as the path to good health. We all need exercise, fresh air, sunshine, enough sleep, recreation, and sufficient self-discipline to stop and smell the roses now and then. Of all the varied professions in the healing arts, including all the alternative healing arts, none of them can heal damaged tissue without nutrition. Certainly drugs can't build new tissue. Nor can polarity therapy, acupuncture, or herbs.

BIOCHEMICAL INDIVIDUALITY

Only foods provide the exact nutrients the body needs to heal itself. Different persons may need different amounts of particular vitamins, minerals, trace elements, protein, carbohydrates, fats, and other nutrients than other people need. RDIs (reference daily intakes) have been applied as general guidelines, but what we all really need is an individual, tailor-made set of vitamin and mineral guidelines that would fit us perfectly.

Forty years ago, Dr. Roger Williams, a University of Texas biochemist, authored a book titled *Biochemical Individuality,* which brings an important principle to our attention. Each of us is unique, and our nutritional needs differ from person to person. The official standards—RDAs (recommended dietary allowances), RDIs (reference daily intakes), and DVs (daily values)—are only best estimates based on averages over large numbers of people. They don't actually fit any person, but they are close enough to help most people use them for guidance. What we really need is a relatively precise assessment of our individual nutrient needs by nutritionists or other food-wise professionals who know how to calculate the exact amounts of vitamins and minerals to compensate for genetic weaknesses and enzyme support and who can identify excess intake of other vitamins and minerals that we need to cut back. Perhaps this kind of advice will become available at a reasonable cost in the healing arts of the near future.

The highest ideal of the sincere health professional should be the well-being of his or her patients, which would seem to involve selecting the least invasive treatment, the option that would cause the fewest side effects and offer the least risk of future problems. There is only one remedy that fits that description for most people with health problems: whole, pure, and natural food, as fresh as you can get it, prepared in a manner that preserves as much of its nutritional value as possible.

Drugs are simpler and more convenient to administer, but they all initiate unnatural chemical reactions or harmful biological reactions (especially steroids and antibiotics) in the body. I am not opposed to using drugs to save lives when there is no effective alternative therapy, but everyone who uses drugs—prescription, over-the-counter, or street drugs—is introducing chemicals into his or her body that the body was never designed to assimilate or process. Often, the liver enzymes have to break down or detoxify at least part of most drugs. There is a price to be paid by the patient for using drugs, legal or illegal, and that price impacts the person's health and future well-being. I would like to see the health-care profession in the United States have the necessary courage and compassion to make food and appropriate food supplements the therapy of choice for every patient who can possibly-be helped effectively without drugs.

I believe in the natural control of disease whenever possible because so many of the prescriptions and therapies of our time leave the patient at a lower level of health than before treatment. Foods don't do that. I believe that when the microbiology experiments and discoveries of the past twenty years have been assimilated into the medical school curriculum, a revolution is going to take place. I believe that the balancing of body chemistry by selected nutrients will become the obvious and preferable means of restoring the functioning of enzyme reaction sequences, thus bringing about restored function at the level of organs, glands, and tissues.

My work has been the primary motivating factor in my life because I feel so encouraged by the thousands of patients whose lives and vitality have been restored by natural means. Nature always has a remedy, but sometimes she needs a helping hand. I only wish that the long, slow increase of national interest in natural and alternative healing methods demonstrated over the last two decades would suddenly escalate into a tidal wave that would cleanse our nation of the many plagues that beset it in our time.

INDEX

G

H

CPSIA information can be obtained
at www.ICGtesting.com
Printed in the USA
JSHW010724170523
41814JS00009B/414

9 780658 002779